Uses of Plants by the Hidatsas of the Northern Plains

Uses of Plants by the Hidatsas of the Northern Plains

GILBERT LIVINGSTON WILSON
Edited and annotated by Michael Scullin

University of Nebraska Press
Lincoln & London

© 2014 by the Board of Regents of the University of Nebraska

All rights reserved
Manufactured in the United States of America

Library of Congress Cataloging-in-Publication Data

Wilson, Gilbert Livingston, 1868–1930.
Uses of plants by the Hidatsas of the northern plains /
Gilbert Livingston Wilson; edited and annotated by Michael Scullin.
pages   cm

Includes bibliographical references.
ISBN 978-0-8032-4674-4 (cloth: alk. paper)—ISBN 978-0-8032-6775-6 (epub)
ISBN 978-0-8032-6776-3 (mobi)—ISBN 978-0-8032-6774-9 (pdf)
1. Hidatsa Indians—Ethnobotany. 2. Plants, Useful—Great Plains.
3. Hidatsa Indians—Material culture. 4. Hidatsa Indians—Gardening.
5. Indians of North America—Great Plains—Ethnobotany.
6. Ethnobotany—Great Plains. I. Scullin, Michael, editor. II. Title.
E99.H6W755 2014   978.4004'975274—dc23   2014010377

Set in Sabon by Renni Johnson.

CONTENTS

List of Illustrations   xi

Preface   xv

Acknowledgments   xix

Introduction   xxi

Editor's Note   xxxvii

Abbreviations: BBW=Buffalobird-woman;
PW=Poor Wolf; GB=Goodbird;
SW=Sioux Woman; GLW=Gilbert Wilson;
WC=Wolf Chief; MS=Michael Scullin

1. Plants That Are Eaten

Domesticated plants (MS)   3

Sunflowers (BBW)   17

Corn-smut (BBW)   20

Prairie turnips (BBW)   22

Jerusalem artichokes (BBW)   31

Hogpeanut (BBW, WC, GB)   36

Chokecherries (BBW)   43

Buffaloberries (BBW)   50

Gooseberries (BBW)   52

Black currants (BBW)   54

Wild grapes (BBW)   56

2. Plants That Can Be Eaten

Hawthorns (BBW)   59

Wild white onions (BBW)   61

Ball cactus (BBW, WC)   63

3. Plants That Are Sweet

Juneberries (BBW)   71
White juneberries (BBW)   78
Wild plums (BBW)   79
Strawberries (BBW)   83
Roses (BBW)   85
Red raspberries (BBW, SW, GB)   91
Biscuitroot (BBW)   93
Nannyberries (BBW)   97
Purple prairie clover (BBW)   99

4. Plants That Are Good to Chew

Sticky gum (BBW)   103
Pine pitch (BBW)   105

5. Plants That Smell Good

Purple meadow-rue (BBW)   109
Blue giant hyssop (BBW)   111
Sweetgrass (BBW)   112
Wild bergamot (BBW)   117
Pine needles (BBW)   119
Perfumes used in beds (BBW)   121
Beaver musk (BBW)   123

6. Plants That Have Medicinal Uses

Big medicine (BBW)   127
White and red baneberry (BBW)   128
Gumweed (WC)   130
Purple coneflower (WC)   132
"Medicine in the woods" (BBW)   134
Poison ivy (BBW)   135
Unknown grass (BBW, GB)   137
Peppermint (BBW)   138

7. Plants Used for Fiber

Dogbane (WC)   141
Upright sedge (BBW)   155
Grasswork ornaments on leggings   159

8. Plants Used for Smoking

Tobacco 9a (BBW)   163
Tobacco 9b (WC)   172
Red-osier dogwood (BBW)   187
Bearberry (BBW)   189
Bearberry or kinnikinnick (WC)   191

9. Plants Used for Dye and Coloring

Yellow owl's-clover (BBW)   197
Water smartweed (BBW)   198
Dye plants—unidentified (BBW)   199

10. Plants Used for Toys

Umakixeke, or game of
   throwing sticks (BBW, GB)   203
Popguns (BBW)   204
A toy horse   205
Reed whistle (GB)   206

11. Plants Used for Utilitarian Purposes

Cordgrass (BBW)   211
Buckbrush (BBW)   213
Cattails (BBW)   219
Boxelder (BBW)   222
Buffalograss (BBW)   226
Big bluestem (WC)   228
Common rush (BBW)   231
Scouringrush horsetail (WC)   237
Puffball (BBW)   239

Snakewood (BBW, WC)   241
Goldenrod (BBW)   244
Prairie grasses as fodder (WC)   246

**12. Plants Used for Rituals or
with Ritual Significance**
The three kinds of sage (WC)   251
Pasture sage 1 (BBW, GB)   256
Pasture sage 2 (BBW, WC)   258
Common sagewort (BBW, WC, GB)   261
Black sage (BBW, WC)   262
Fringed sage (PW)   268
Juniper (Cedar) (BBW, WC, GB)   269
Creeping juniper (BBW, GB)   270
Prairie sandreed (WC)   271
Bittersweet (WC)   275

13. Sources of Wood
Wood as a resource (MS)   279
Cottonwood (WC)   284
Ash (BBW)   289
Peachleaf willow (BBW)   291
Sandbar willow (BBW, WC, GB)   294
Heart-leaved willow (BBW)   297
Quaking aspen (BBW)   299
American elm (BBW)   300
Water birch (BBW)   301
Boxelder (BBW)   302

14. Uses of Wood
Gathering firewood (WC)   305
Digging-sticks (BBW, WC)   311
Mortar and pestle (BBW)   314
Making a bullboat frame (BBW)   316
Making a wooden bowl (WC)   320

Rakes (and the bison scapula hoe)
(BBW, WC)   325
Paddle for working clay pots
(cottonwood bark) (GLW)   329

15. Arrows
Significance and utility (MS)   333
Making arrows (WC)   335
Types of arrows (WC)   344
Bows (WC)   347
Arrows for boys (BBW, GB)   350
Mock battle with grass arrows (WC)   354

16. Earthlodges
Building an earthlodge (BBW)   359
On Earthlodges (The observations of
Hairy Coat and Not A Woman)   370
Winter lodges and twin lodges (BBW)   374
The peaked or tipi-shaped
hunting lodge (BBW)   378
The use of sod as an earthlodge covering   382
Dismantling an old earthlodge (BBW)   384
Like-a-Fishhook Village and
environs (WC)   389

17. Miscellaneous Material
Basket making (BBW)   395
Native drinks of the Hidatsas (BBW)   403
How our meals were served (GB)   406
Nettles (BBW)   409
Forest fire (GLW)   411

Conclusion   413
Appendix: Frederick N. Wilson's
Comments on "The Hidatsa Earthlodge"   419
Bibliography   427

ILLUSTRATIONS

MAP

1. Great Plains prairie types and location
   of Fort Berthold   xxx

FIGURES

1. Son Of A Star and Buffalobird-woman's winter
   and summer homes in 1878   xxvii
2. A mouse-proof hanging sling   47
3. Woman playing a game   51
4. A currant   52
5. Wild white onion   62
6. A smaller mouse-proof sling   75
7. A sage-lined pit for ripening plums   80
8. Hidatsa gaming pieces made from plum pits   81
9. Mandan gaming pieces made from plum pits   82
10. A scrotum basket   86
11. "Wild carrot" (biscuitroot)   94
12. Sweetgrass braids on an eagle feather fan   113
13. Sweetgrass tie on Buffalobird-woman's braid   115
14. Pine needle necklace for a child   120
15. Bed frame   122
16. Location of beaver scent glands   124
17. A model of a deer snare   145
18. A rabbit snare   147
19. Carrying a beaver grass bundle   156
20. Needle for stringing dried squash,
    with a beaver grass cord   157
21. A bone awl   160

22. A tobacco blossom with calyx   173

23. Drying and storing tobacco   174

24. A reed whistle   206

25. The other reed whistle   206

26. Buckbrush bundle   214

27. A snare fence of buckbrush   216

28. An improvised sweatlodge frame   217

29. Collecting boxelder sap   223

30. A boxelder root hanger   224

31. A red-grass (big bluestem) arrow   230

32. The "ornamented cradle"—the model
for the reed doll   235

33. Ghost whistle   238

34. A sandreed (*Calamovilfa*) hair ornament   272

35. Wolf Chief with his hair ornament   273

36. Wood framing of an earthlodge   282

37. Willow fence showing details of construction   298

38. Buffalobird-woman with a load of firewood   306

39. Small Ankle snagging driftwood from the Missouri   307

40. Buffalobird-woman carrying a log   308

41. The basic garden/all-purpose ash digging-stick   312

42. A bullboat frame   317

43. A large wooden bowl with a lug or handle   322

44. A burl on a cottonwood tree   323

45. A wooden cup   324

46. A bison shoulder blade (scapula) hoe   325

47. A wooden ash rake and an antler rake   326

48. The "pottery patter" of cottonwood bark   330

49. Three types of arrows   344

50. An unstrung bow in its case   345

51. Setting the curve in a bow   348

52. A boy with his grass arrows, quiver, and bow   351

53. Boys hunting with blunt arrows   353

54. Buffalobird-woman measuring lodge post placement   363

55. Placement of a vertical center post   364
56. Raising a horizontal stringer   364
57. Two horizontal stringers on a vertical post   365
58. The framing of a 13-post lodge (from above)   366
59. The framing of a 13-post lodge (side view)   367
60. Winter earthlodge framing   375
61. Structure of a "twin-lodge"   376
62. Supporting posts on hunting lodge   377
63. Details of the construction of a hunting lodge roof   380
64. Post-pulling device   388
65. Design of Like-a-Fishhook Village   390
66. Like-a-Fishhook Village and its environs   391
67. The palisade, with a loophole and ditch inside   392
68. A bastion in the palisade   392
69. A basic basket pattern   398
70. Wooden bowl with horn spoon   404
71. Goodbird eating with his parents in hunting camp   408

PREFACE

In old times our tribe knew many plants and ate them each
in its proper season, for they knew when each should be
gathered. And because they knew each plant's proper season,
no one got sick and no harm came of eating them.
Such was our custom.

In old days, also, we had no whitemen with us and no guns,
and we often faced scarcity of food. And so we ate such
foods as wild carrots. But in my day it was not necessary to
do this, and such wild foods fell into comparative disuse.

—BUFFALOBIRD-WOMAN

Gilbert Livingston Wilson (1869–1930) was, in many respects,
an anthropologist well ahead of his time. Anthropology in the
first few decades of the twentieth century was highly focused on
collecting any cultural information that could be salvaged from
American Indians who had lived in and survived the nineteenth
century. Of most interest to anthropologists of the era was recording the variations between groups in matters of social organization, religious practice, and material culture—an impersonal or
detached view or what some today would call an "etic" perspective (Pike 1967). Wilson, in contrast, was concerned with what
he saw as the failure of anthropologists to value the perspectives
of individual members of a culture—to allow these individuals
to tell their own stories—and to understand how individuals
understood themselves, their surroundings, and their cultures.

This, he believed, allowed for the transmission of the richness and depth of personal perception. This would be an "emic" perception (Pike 1967). In a letter to Clark Wissler (May 6, 1913: 2–3) Wilson wrote:

> I do not know that I have said it before, but my reports, as they are sent in to the museum, are wrought out on a plan which I have purposely chosen. Most of the papers put out by the Museum and other scientific bodies seem by me to lack one thing that to an amateur student might be of value, tho I do not mean to say that as a criticism. But to a student of the Indians who relies wholly on books, there is a difficulty in getting the proper conception of Indian character. While only living acquaintance can do this satisfactorily, it has seemed to me that a report can do much to assist in this important part of the student's investigations, if the report itself is given as the Indian gave it himself in taking dictation. In taking dictation I keep this in mind. I take down every word that is said; when rewritten in typewritten form, the matter is merely rearranged, and if the informant has covered the same ground twice, I merely combine the two paragraphs in one without really altering the language.
>
> My aim is, if death or disability removed me from doing anything further with the reports myself, that any student who studies them will find the Indian's way of looking at things to be retained as nearly as possible in another language.

Wissler, who was curator of anthropology at the American Museum of Natural History and who had hired Wilson in 1908 to collect both ethnographic data and specimens for the museum's collections, took offense and accused Wilson of having "contempt" for the largely traditional ethnographic methods employed by himself and other anthropologists working for the museum, including Robert Lowie, who was at Fort Berthold for a little while when Wilson was working there and who interviewed some of the people with whom Wilson had worked (e.g., Lowie 1917). Wilson hastened to assure Wissler that he held no contempt for traditional anthropological methodologies, but that first and foremost he wanted the Indians to speak for themselves (Wissler 1916; Wilson 1916).

Wilson's dissertation, "Agriculture of the Hidatsa Indians: An Indian Perspective," was completed in 1916 and published the following year. It is currently in print under several titles, usually "Buffalo Bird Woman's Garden." This dissertation on Hidatsa gardening as recalled by Buffalobird-woman is both highly autobiographical and individualistic. It has become an anthropological classic as perhaps the first reasonably autobiographical account by an American Indian woman and because of the great attention given to the details of Hidatsa gardening from the perspective of the gardener. That Wilson insisted on writing in a highly accessible style makes all of his publications both readable and memorable.

ACKNOWLEDGMENTS

Tom Thiessen read through the manuscript and, noting that he had a few comments, went on to fill numerous pages with good advice of both the obvious (which had escaped me) and the obscure (about which I had never thought).

Kristen Mable at the American Museum of Natural History graciously found and copied material from the museum's anthropological archives, and I appreciate the extra time and effort she put into this task.

Most of all, my wife, Wendy, was a constant. I would disappear into the keyboard and emerge with various consternations and she, one way or another, was there to get me back on track. She read through the entire manuscript and pointed out multitudinous errors of spelling, grammar, and organization. She has been a most patient editor as time warped by. I dedicate this book to her.

INTRODUCTION

Gilbert L. Wilson's ethnobotanical project had its origins in his collaboration with Professor Josephine Tilden of the Botany Department at the University of Minnesota. At her behest, in the summer of 1916 Wilson collected examples of plants used by the Hidatsas and the purposes for which they had been used. Tilden and her students planned to identify them and compile the information into a text. Although Tilden did send a student along with Wilson (a man identified only as "Haupt") to do some of the collection and field preparation of specimens, for unknown reasons the project was never completed and the plant specimens have been lost. Wilson, for his part, recorded what Buffalobird-woman and her brother, Wolf Chief, had to say about various plants and their uses. Their recollections and associations make this text unlike any ethnobotanical monograph.

Wilson had already spent parts of ten summers at Fort Berthold Reservation in west-central North Dakota, almost all of that time with Buffalobird-woman, Wolf Chief, and Buffalobird-woman's son, Goodbird, acting as interpreter. Wilson also became an adopted son of Buffalobird-woman, and thus they were all members of the Prairie Chicken clan.

The accounts of different plants tend to vary immensely according to the significance of the plants to the speaker and according to their recollections of them. Wilson apparently had no particular agenda other than to write down what they had to say about whatever plants they thought important enough to mention. I have added information on plants that Wilson had transcribed in other years and a section on wood. Wood was perhaps the most important plant resource to the Hidatsas, one of a number of groups practicing horticulture and bison hunting. All these groups lived in more or less permanent villages of large earthlodges, as opposed to the mobile hunters who lived in tipis.

The Hidatsas (and Mandans and Arikaras) were separated (by government mandate) from their last earthlodge village at Like-a-Fishhook (also known as Fort Berthold) in 1885. Most of Buffalobird-woman's recollections of her gardening days (Wilson 1917a) seem to date to the period between 1845 and 1885 during which time all three tribes lived in Like-a-Fishhook, and much of what she and Wolf Chief have to say about plants also seems to date from that time as well.

In 1916 she was seventy-six or seventy-seven years old and Wolf Chief was sixty-six. This means that more than thirty years had passed between their life in the last earthlodge village and their subsequent life in dispersed individual log cabins on allotted land. By 1910 only one earthlodge, that of Hairy Coat, was still inhabited, although several others were still standing. Thus some recollections are a bit hazy, and in some cases the plants described are unidentified or unidentifiable.

Buffalobird-woman was the most conservative of the three people with whom Wilson worked almost exclusively, and she was the only one who had not converted to Christianity. Furthermore, she was a great stickler for details. She was sometimes reluctant to discuss a plant because it was not hers to talk about, the rights to some plants being owned by individuals or societies. For some plants, such as sunflowers and prairie turnips, Wilson seems to have felt that he already had all the information he needed.

Wolf Chief loved to tell stories and was good at it. More often than not he included anecdotes about how he or others were using or had used a particular plant. In the lengthy section on tobacco, one gets both Buffalobird-woman's perspective on a plant grown and used only by men as well as the perspective of Wolf Chief, who actually maintained tobacco gardens over the years and continued to prepare tobacco and other plants frequently mixed with it for smoking.

Goodbird occasionally added the equivalent of a footnote as well as a few recollections from his childhood.

The results are highly uneven. There is considerable detail on some plants and virtually nothing for others. Wilson never got beyond transcribing his field notes (or having them transcribed, which was usually the case) for that part of his 1916 fieldwork, and he seems to have done no editing of it whatsoever.

World War I kept Wilson from going back to Fort Berthold in 1917 and may well have been the reason the project was never completed. For one thing, the American Museum of Natural History had no money to sponsor his fieldwork in 1917. For another, Wilson, in his capacity as a clergyman, worked very hard on the home front at patriotic rallies and at explaining the war to an eastern Minnesota population that included many recent immigrants from central Europe (see Wilson 1917b).

I could find no information on the fate of the plant specimens collected for the botany project. The plant collection, which apparently made it to the University of Minnesota, has disappeared. A duplicate collection made for the American Museum of Natural History has likewise disappeared. Thus the final identification of some plants is in question because I had nothing to work from other than Wilson's notes and Goodbird's drawings of some plants. And, of course, Wilson was not a botanist and knew little about plants.

In 1918 Wilson went to North Dakota one last time, with the intention of wrapping up eleven years of fieldwork at Fort Berthold Indian Reservation. But because his health was failing, he never returned to Fort Berthold, and although he did publish one more major monograph in 1924, *The Horse and the Dog in Hidatsa Culture*, his plans to publish much of the rest of his data were never realized.

Buffalobird-woman, Wolf Chief, and Goodbird

Wilson first visited Fort Berthold in the summer of 1906, at which time he was introduced to a Hidatsa named Edward Goodbird. Goodbird knew some English and was pastor of a small congregation. Wilson, at that time a Presbyterian minister in Mandan, North Dakota, was an avocational anthropologist and archaeologist. He and Goodbird soon became good friends and collaborators. Both were born in 1869. Goodbird's parents were Buffalobird-woman and Son Of A Star, who died not long after Wilson met him.

Wilson found Buffalobird-woman (Maxidiwiac, in Hidatsa) to be an ideal source of information about Hidatsa life. She was about sixty-seven (having been born in 1839 or 1840) when he met her, and she had lived for forty-five years in earthlodge villages—forty years at Like-a-Fishhook Village, the last of the earthlodge villages.

Her brother, Wolf Chief, was born in 1849. He enjoyed talking about his life and had lived through many of the often terrible transitions of the mid-nineteenth century. Wolf Chief became Wilson's principal source for male perspectives on Hidatsa life.

A man between cultures, Wolf Chief grew up in the presence of "the whitemen," and they were very much a part of his life—sometimes a help and sometimes a hindrance. At age thirty he decided that he needed to learn English. To this end he enlisted a non-Indian friend as tutor and then for a while attended school with the reservation children. He never learned to speak English, but he did learn to read and write it.

Edward Goodbird was the pastor of the Independence Congregational Church (Independence being the principal town on the reservation at that time). Wilson, at times, both lived in and used Goodbird's church as a field lab.

Goodbird became his most valuable ally, translator, diplomat, source of information about the reservation and its inhabitants, and illustrator of many facets of Hidatsa life. Goodbird's English was not fluent, and Wilson once expressed some consternation over words Goodbird had picked up from the soldiers at Fort Berthold, which was immediately adjacent to Like-a-Fishhook. But Wilson also had this to say about Goodbird (vol. 27, 1918: 101–2):

> During the ten or twelve years of the author's work among the Hidatsa, Edward Goodbird was his faithful and efficient interpreter. Goodbird's English education is meager, but he is a natural scholar, and speaks the Dakota, Mandan, and Crow languages besides Hidatsa and English.
>
> The author soon found that Goodbird's rather broken English was a distinct advantage, since it translated so accurately the thought and form of the Hidatsa idiom. The author has recorded the following account exactly as it fell from the interpreter's lips.

Although Wilson did on occasion work with other Hidatsas and Mandans, he spent by far the largest amount of his time with Buffalobird-woman, Goodbird, and Wolf Chief. The concept of sampling or cross-checking information was seldom important to him, so virtually all the information gathered for this volume is from Buffalobird-woman and most of the remainder from Wolf Chief. But Wilson was a busy man, and he had many other things to accomplish in that summer of 1916. He spent about two weeks on the project and collected data, some extensive and some sketchy, on almost eighty species of plants.

Some Observations on Hidatsa Gardening

Although this book is about the Hidatsas' uses of plants, little is said about Hidatsa horticultural practices, although gardening was certainly their most important use of plants, as far as subsistence was concerned. The reason for this is that many of Wilson's interviews with Buffalobird-woman (particularly those of 1912) were to became his dissertation in anthropology. It was first published in 1917 by the University of Minnesota as "*Agriculture of the Hidatsa Indians: An Indian Interpretation.*" In my opinion no better book has ever been published on the topic. Buffalobird-woman was an ideal person to be Wilson's source, and Wilson was the right person to put it all together in a readable and accessible form. Anyone wanting information about Hidatsa gardens and gardeners can easily find Wilson's dissertation, now entitled "Buffalo Bird Woman's Garden" and published by at least three presses. There is also an online version available from the library at the University of Pennsylvania.

Buffalobird-woman was part of the second generation of Hidatsas to have fairly extensive contact with Euro-Americans and the first generation to be born after the horrors of the smallpox epidemic of 1837, which decimated the Hidatsa people. The changes in Hidatsa culture that occurred during her lifetime are truly mindboggling. Her family went from using the bison scapula hoe of her grandmother to the iron hoes that she grew up with and were considered the standard garden implement. In later years Wolf Chief used a horse-drawn plow for the garden. Despite the epidemics, despite the warfare with the Lakotas and the Assiniboines, despite losing her first husband to disease, and despite being evicted (along with everyone else) by the U.S. government from her home in Like-a-Fishhook Village in 1885, Buffalobird-woman still held on to her memories of a happier past ("the old days"). (See Smith 1972 for a concise history and the archaeology of Fort Berthold/Like-a-Fishhook and Meyer 1977 for a detailed history.)

Wilson was dedicated to allowing the people he was interviewing to interpret their own lives; he listened and transcribed. Not that he didn't ask questions or try to keep the conversation firmly on the course he planned, but he was decidedly not an orthodox anthropologist of the second decade of the twentieth century. His training as a Presbyterian minister did nothing to prepare him to understand the functioning of a garden and the

1. In 1878 Buffalobird-woman and her husband, Son Of A Star, lived in the log cabin in the winter and in the earthlodge the rest of the year. Drawing by Goodbird. (Courtesy of the American Museum of Natural History)

tribulations of gardening. North and South Dakota are notorious for extreme weather: killer blizzards, blazing drought, and horrific thunderstorms with great blasts of wind and hail that pulverize everything to an unrecognizable green pulp. Summer high temperatures can easily exceed 100 degrees (38 C) at which temperature, as I have observed in our own garden, corn cannot function properly and tassels eventually "fire," producing no pollen. Therefore, no kernels will develop on the ears—if the plants survive. And sometimes they don't. But Wilson didn't ask, and Buffalobird-woman didn't volunteer anything about such matters. The great droughts of the mid-1860s (NOAA 2009 and Cleaveland 1999), for example, were not to be recalled.

There were no crop failures to recount nor reasons to explain such failures. If there were garden pests like grasshoppers, raccoons, deer, beaver, gophers, or moles (and there were, you can be sure), she left them out of her recollections. What was the yield of a Hidatsa garden? How much Hidatsa subsistence was derived from garden produce, and how much from hunting and the gathering of uncultivated plants? At the time Wilson was there, some gardens were still being planted and harvested. But while he mapped several, he did not think (or was not inclined) to ask about yield or how reliable the gardeners thought their gardens were.

Having grown Hidatsa corn, beans, and squash for more than thirty years, and having tried to follow as closely as possible the dictates of Buffalobird-woman, I have had questions, and many of them have been either answered or found out (Munson-Scullin and Scullin 2005). But the question still remains: how important were gathered plants to the Hidatsa diet? The answer is that there is no single figure or conclusion to be drawn. Some years the garden would have been a failure or largely a failure, and reliance on bison and perhaps intensified gathering might have been necessary. Other years would have been great successes, but a gardener couldn't count on it—only just hope or pray for it. An old German farmer fishing in the Knife River at the site of the former Hidatsa village of Awatixa said it all when, a few years ago, I commented on how good the crops looked in late July: "It ain't in the bin yet."

Hidatsa Cultural Geography

Hidatsa lives were focused on several variables other than the universals of food, shelter, and security. Because they were village dwellers, they needed a place that had the resources to sustain a village—for years, decades, or even centuries in some cases. Because they practiced both bison hunting and horticulture, the Hidatsas needed a place with access to areas frequented by bison as well as good land that could be tilled with digging-sticks and bison scapula hoes and that could produce at least an adequate crop of corn most years. They also needed trees.

Their dwellings were large and constructed of wood, and their villages were protected by palisades of timber. Virtually all their food needed to be cooked. They therefore required a large and reliable source of wood all year long, and year after year, within an ecosystem where wooded areas were almost entirely limited to the floodplains of the Missouri River and its larger tributaries. There the gallery forests could be cut for timber, and the annual floods brought large quantities of driftwood for both timber and firewood.

Climatic variability affected both the yield of bison and the yield of corn—a precarious set of circumstances. Central North Dakota is in the transition zone from the long grass prairies in the east to the short grass prairies to the west—an area of "midgrass" or "mixed-grass" prairies. This means not only a mixture of tallgrasses (e.g., big bluestem) and short (e.g., buffalograss), depending on the nature of a particular place but also an abundance of midsized grasses such as the needle grasses, grama grasses, and western wheatgrass, which all grow to a foot or two tall (30-60 cm).

Historic records from the past 150 years and prehistoric records from dendroclimatological sources and cores from the bottoms of North Dakota lakes extend the climatic record back at least two millennia (e.g., Laird et al. 1996) and show the alternation between drought and ample to even abundant precipitation varying considerably over time. (For a succinct review of the literature, see Ashworth 1999.) People in the area never knew just what to expect, with the result that climate-mediating religious ceremonies were well integrated into their lives.

Most precipitation in the vicinity of Fort Berthold Reservation falls during the growing season (High Plains Regional Climate Center 2012), but because much of it is derived from thunderstorms, it is literally "hit and miss." Furthermore these storms may produce both high, even violent, winds and hail, which can quickly strip leaves from plants or even destroy them completely. Rivers and streams draining into the Missouri River (almost all from the higher elevations of the Rocky Mountains) may be raging torrents or trickles. It is a land of frustration and uncertainties, which requires diverse and flexible approaches to survival.

Bison, which once numbered in the tens of millions in the Great Plains (estimates vary from 30 to 60 million, e.g., Knapp

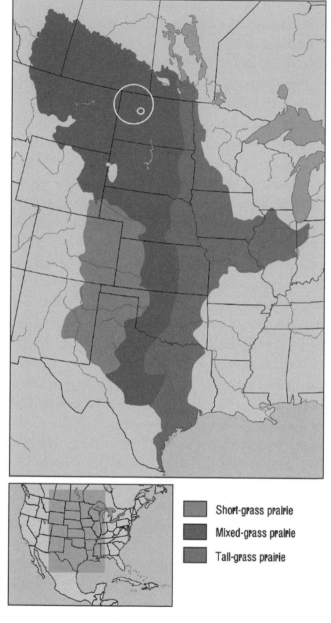

1. Great Plains prairie types and location of Fort Berthold. (Adapted from the original courtesy of James Mason and the Great Plains Nature Center, Wichita KS)

et al. 1990: 39), were the Hidatsas' main source of food and also a valuable source of hides and bones for tools such as the bison scapula hoe, the horticultural standard. Larger bones like the femur and humerus were pounded to fragments with stone hammers and then boiled for the much valued "bone-grease." The fat accumulated in the hump in the fall as well as kidney fat were also much appreciated, but the bone-grease was considered to be the most desired and well worth the extra work required to process it. Any habitation site found in the range of the bison will have shattered bison bone, usually in considerable quantities, in the site matrix and garbage middens.

The North Dakota Badlands, as they are now known, are the deeply dissected soft rocks (sandstones, shales, clays, and a very soft coal known as lignite) exposed on both sides of the Little Missouri River in far western North Dakota. These were deposited as the result of the formation and subsequent erosion of what are now the Rocky Mountains, a process begun about 65 million years ago (see KellerLynn 2007 for a thorough account). These were indeed "badlands" for nineteenth-century cross-country travelers with teams and wagons, but to the people who had lived in or near the area for more than 10,000 years, they were a highly favored hunting territory. Vegetation changes abruptly at the eastern edge—the land literally drops away, sometimes for hundreds of feet—and the various weathering sands and clays support a diverse plant cover. This is, for the most part, short-grass prairie with a distinctly different character than the lands around the villages on or near the Missouri River.

Thus the Hidatsa had a minimum of seven ecosystems from which to extract resources: the Missouri River provided virtually all needed water, driftwood for construction and firewood, and fish; the floodplain adjacent to the river provided habitat for game, trees for construction and firewood, and land for gardens; the high, grass-covered terraces along the Missouri and Knife Rivers were the sites of most villages and, from the early nineteenth century on, pasture for their horses; the wooded coulees (ravines) that drained the higher lands into which the Missouri River was entrenched provided sites for the smaller winter villages because they offered shelter from the winds, wood for dwellings and fuel, and a spring for a reliable source of water; the wet or at least damp areas of scattered springs, sloughs, and creeks provided game and the grasses used in the construction of earthlodge roofs and the linings of storage pits; the vast mid-grass prairies supported what seemed to be an infinite supply of bison; and the frequently visited badlands were populated by bison, mule-deer, elk, antelope, and smaller game. The badlands also provided a considerable diversity of plant communities that were dependent on both soil types and aspect—the direction faced by the slope that determines whether the slope is hot and dry (south and southwest facing) and therefore supports a cover of grasses and drought-resistant forbs and shrubs, or cooler and moist enough (north and northeast facing) to support woody vegetation like junipers, ash, chokecherries, and juneberries as well as a diverse understory.

The climatic extremes of the plains meant that failure of one or more of these systems was inevitable at any unpredictable time. People who lived in these high-risk areas had multiple alternative strategies for coping with the various adversities with which they were confronted. Knowing the land and its resources, as well as its great capacity for unreliability, was the highest priority for those who lived there.

The Hidatsas, the Mandans, and the Arikaras

The MHA (Mandan, Hidatsa, and Arikara) Nation, or the Three Affiliated Tribes, as they are known today, were all village dwellers with economies based on both horticulture and bison hunting. To call them "horticultural tribes" is to place undue emphasis on the role of horticulture in their subsistence. Theirs were very much mixed economies with the fraction of their subsistence from gardening and from bison hunting varying considerably from year to year as well as from family to family.

The Mandans and the Hidatsas speak languages of the Siouan Language Family, and at various times in the preceding thousand years the past thousand years they have arrived at the Missouri River in what is now North Dakota from the east and southeast. They brought with them eastern horticultural traditions and village life. The Arikaras are a branch of the Pawnee (Caddoan Language Family) who occupied the central Great Plains. The Arikaras made their way up the Missouri River in late prehistoric times until all three tribes wound up living in a single, well-fortified village in the mid- to late nineteenth century. This village was Like-a-Fishhook. Their histories and prehistories, needless to say, are long and complex, but for the Hidatsas the best and most accessible source is *People of the Willows* (Ahler, Thiessen, and Trimble 1991). *The Way to Independence* by Carolyn Gilman and Mary Jane Schneider (1987) is no longer in print but is an extraordinary book that provides many details of the lives of Buffalobird-woman, Wolf Chief, and Goodbird, as well as the best biography of Gilbert Wilson ever produced (thanks to Alan R. Woolworth). The book, which contains hundreds of illustrations, is a veritable catalog of the material culture of the era. *The Village Indians of the Upper Missouri* by Roy Meyer (1977) is a good history of all three tribes. Volume 13 of the *Handbook of North American Indians* (DeMallie 2001) provides encyclopedic coverage of Plains Indian cultures from prehistoric through historic times.

Some Observations on Indigenous Horticulture

Once people have made the shift to growing at least some of their own food, they are compelled by their gardens to settle down. Gardens require frequent, if not almost constant, attention in order to function properly. As I have frequently noted, nature abhors a garden.

Gardening involves location of land appropriate for gardens; clearing land to prepare it; tilling the land with digging-sticks (until horses were put to that enormous labor in the latter part of the nineteenth century) so that planting is possible; weeding the crops so that plant competition is kept under control; protecting them from the depredations of wild animals like deer, raccoons, and beaver as well as other humans; harvesting; and storing most of the harvest for later consumption. All of this demands that the gardeners be there a great deal of time. Furthermore, after all of this, the produce must be processed to make it edible.

Harvesting may well mean that a ton or more of processed crops have to be stored securely so that it is available when needed as well as safe from those who would steal it—always a serious problem. To protect their harvests, people dug pits that were lined with grass and then thoroughly obscured at the surface so that only family members would know the location come winter when the cache was opened and food taken to the sites where people overwintered.

During the nineteenth century, new garden plants were often quickly adopted. A long grown variety of *Cucurbita pepo* was replaced by other varieties—like the pat-a-pan and new species such as various varieties of Hubbard squash (*C. maxima*), which has both a thicker flesh and a sweeter taste. Different varieties of beans grown by the Hidatsas, Mandans, and Arikaras could compete with the "red bean" still being grown today. The yellow bean in particular seems to thrive in hot, dry weather and probably originated in the Southwest among the Pueblo peoples. Shield beans are larger, take longer to mature, and are widely distributed, at least over the eastern half of the United States.

Watermelons (*Citrullus lanatus*) are an example of plants that were rapidly integrated into gardens. They were introduced after having rather quickly made their way from the southeast Atlantic coast, where they had been introduced from Africa by European colonists (Blake 1981). "Spanish settlers were producing watermelons in Florida by 1576" (Maynard and Maynard 2000), and they had reached the middle reaches of the Missouri River by the early 1800s, as noted by William Clark on August 2, 1804. Why did watermelons travel so quickly? Because they are sweet and because they will grow under drought conditions with which other garden plants struggle. They also produce numerous seeds.

Gardeners are usually quite happy to try new plants if they are tasty or in some way useful, but if they are sweet, they are even more welcome. Wild plants, in the grander scheme of things, were relatively unimportant to regular diets. As food they were largely complementary unless a person or group was starving.

On the other hand, gathered plants offered very utilitarian raw materials to produce fiber for twine, pigments for facial and body decoration, medicine, and perfumes. A few plants, like the sages and junipers, had ritual uses. Many gathered plants, however, added an important but missing ingredient: something sweet.

Wood was a special category of "gathered plants" but was ubiquitous in the lives of the Hidatsas. They lived in wooden lodges, which might require two hundred or more trees of varying size. Their villages were protected by wooden palisades, which used many hundreds of trees. Hunting and gardening tools, bullboats, and bows and arrows were constructed of wood. Each village consumed tons of wood every year, and the quest for wood was a constant in life.

Knowledge of plants was widely shared. The late prehistoric and early historic time frame during which many peoples existed on the Great Plains was of relatively short duration—a few generations at most for some—and therefore each newly arrived group had to tap into the existing knowledge base of those who had lived there long enough to develop an extensive familiarity with the local flora.

People trade information as well as tangible items. In fact, the trade of tangible items is invariably the occasion for the transfer of information as well. If there were not sufficient reasons to establish trading relationships, then people created them. Napoleon Chagnon's descriptions of Yanomamo trading and alliance formation were certainly mirrored on the plains (Chagnon 1997: 162–64). Raymond Wood (1980: 108) noted that on the Great Plains an important "consequence of systematic trade . . . was the diffusion of cultural elements over large areas. This contributed to cultural uniformity among alien groups and provided a means of leveling cultural differences over wide areas."

People have long favored having trading partners, people they hope that they can trust. On the plains (as elsewhere) this relationship was often strengthened by extending kinship terms to trading partners (brother, father, sister, and so on). Information about plants must have been exchanged as part of the whole process of trading (which sometimes included plants and the rights to use certain plants). Hidatsa and Mandan villages were important trading centers, and hunting people from all over the Northern and Central Plains came there on a regular basis to exchange dried meat, hides, and exotic items for corn, beans, and tobacco. Historically guns obtained from the trappers and traders to the north and east were traded to southern peoples for horses brought up from the south (Lehmer 1971b).

For all the diversity of the origins of tribal groups who came to inhabit the Great Plains in the past thousand years, the area experienced a great deal of convergence of culture and material culture as people quickly learned how to live there and exploit the available resources. As noted, an important part of this convergence involved the transmission of information about the utility (or hazard) of various plants.

EDITOR'S NOTE

I have retained Gilbert Wilson's writing and philosophy throughout. Although some minor editing has been necessary, each account
is essentially presented as Wilson recorded it. I have relied on two
sources. One is the material in his notebooks, which were transcribed on the spot from comments made by Buffalobird-woman
and Wolf Chief and translated by Goodbird. These three were
Wilson's principal Hidatsa consultants. The second source is
the transcribed version he prepared for the American Museum
of Natural History.

Wilson's notes and copies of the reports he sent to the American Museum of Natural History are currently in the collections
of the Minnesota State Historical Society (Gilbert L. and Frederick N. Wilson Papers, vols. 3–38).

The Division of Anthropology Archives at the American
Museum of Natural History and the Minnesota Historical Society have joint responsibilities for the entire collection, which also
includes many drawings and photographs. Wilson's field notes
and his transcriptions vary little, but on occasion I have chosen
one or the other because I think it better suits this collection.

Alan R. Woolworth has written an outstanding account of both
Wilson and his brother, Frederick, in "The Contributions of the
Wilsons to the Study of the Hidatsa." This can be found in Gil-
man and Schneider's *The Way to Independence* (1987: 340–47).
For those whose interests in plants extend beyond the material
collected from Buffalobird-woman, Kelly Kindscher's two books,
Edible Wild Plants of the Prairie (1987) and *Medicinal Wild Plants
of the Prairie* (1992), should be consulted. Melvin Gilmore's clas-

sic *Uses of Plants by the Indians of the Missouri River Region* was published in 1919 at about the same time that Wilson was finishing his fieldwork at Fort Berthold. *Montana Native Plants and Early Peoples* by Jeff Hart (1976/1996) discusses many of the same plants encountered by the Hidatsas. All these books are still in print and readily available. Dan Moerman at the University of Michigan has compiled an immense database, *Native American Ethnobotany,* available at http://herb.umd.umich.edu/ and also published *Native American Ethnobotany* (1998).

For some, but not all, of the plants in the text I have added some prefatory remarks. On occasion, comments are inserted into Wilson's transcription within brackets. Wilson also added some explanatory notes or issues, and these are usually contained within parentheses and generally followed with his initials, GLW. For clarity's sake, some parenthetical material is unattributed and may be Wilson's, Buffalobird-woman's, or Wolf Chief's. The source indicated after the name of the individual providing the information or anecdote (e.g., Buffalobird-woman [vol. 20, 1916: 184–87]) indicates the Minnesota Historical Society microfilm number of Wilson's archives.

The book is organized according to the use or uses of the plants as described by Buffalobird-woman and Wolf Chief. A plant's common name is followed by its use or uses in parentheses. The chosen common name and the essential utility of a particular plant are subjective. Common names vary immensely from one part of the country to another, as well as from one user to another. Generally the most frequently used common name is used. Other common names are sometimes provided at the end of each description in the taxonomic section, the organization of which is that used by the U.S. Department of Agriculture in the online plants database (USDA, NRCS 2013; PLANTS Database [USDA 1913]). Because Buffalobird-woman and Wilson were sometimes not explicit in the primary uses of some plants, an attempt was made to determine their use.

The spellings of Hidatsa words appear as Wilson spelled them. He changed spellings as he learned to better understand the Hidatsa language, so many spelling variations exist in his notes from year to year. Once his fieldwork was done, he did a lot of thinking about how Hidatsa words should be written. During this time he also decided that Maxidiwiac's name should be written in English as Buffalobird-woman (Cowbird-woman). Although he did use the spelling Buffalobird-woman in *Agriculture of the Hidatsa Indians* (Wilson 1917a), he and various publishers have used a number of other spellings.

Also included is a fairly detailed account of gardening and the various plants harvested. Because the most important plants consumed were those grown in the gardens, this section is intended to augment the detailed accounts of Buffalobird-woman (Wilson 1917a).

Buffalobird-woman at one point observed:

> I have heard that the Sioux and other tribes gathered wild plums and dried them, but we Hidatsas never did. We raised corn and vegetables, but other tribes had to gather wild fruits not having our garden foods. Having plenty of foods ourselves, we did not have to gather inferior wild foods to store away (vol. 20, 1916: 261–63).

ONE

Plants That
Are Eaten

Domesticated plants

Michael Scullin

A book that is, in many respects, an ethnobotanical treatise must, I think, deal with those plants the author or editor considers to be the most important: plants that provided wood and plants that were grown in gardens. Buffalobird-woman and Gilbert Wilson have given us an extraordinarily detailed account of Hidatsa gardening in *Agriculture of the Hidatsa Indians: An Indian Interpretation* (1917). But there is more to it than Buffalobird-woman managed to recall or that Wilson was able to record. Therefore the following account includes some of what I have learned over the past thirty years and some of what my wife, Wendy Munson Scullin, and I have learned under conditions as controlled as one can have in a garden such as the Hidatsas maintained.

Garden plants played a far more important role in Hidatsa subsistence than did wild plants and had been more significant even before the Hidatsas and their neighbors, the Mandans, arrived in what is known to anthropologists and archaeologists as the Middle Missouri (that part of the Missouri River flowing through North Dakota and South Dakota). Wilson's dissertation for the Anthropology Department at the University of Minnesota (awarded in 1916 and published in 1917) later was published by the Minnesota Historical Society as *Buffalo Bird Woman's Garden* (1987), which is still in print and still popular.

Although it would seem that just about every possible aspect of gardening was covered in Wilson's dissertation, it turns out that Wilson failed to ask some questions that would be fairly obvious to a gardener but were not to the Presbyterian minister turned anthropologist. Many gardeners, but not Buffalobird-woman, would have thought to include things that gardeners are always talking about: the weather, the bugs, the weeds, the yields. But she didn't. Buffalobird-woman's memory was suffused with a golden glow that managed to obscure epidemics, wars, droughts, hail-storms, grasshoppers, strong winds, aphids, and assorted "gar-den variety" plagues that usually leave all-too-vivid memories.

In my more than thirty years of growing the garden described by Buffalobird-woman, I have learned a lot that she never dis-cussed. I have been (repeatedly) struck by the almost never-ending problems and the sometimes extraordinary variability of our corn production from year to year ("corn" being used interchangeably with the more correct term, "maize"). My wife and I have mea-sured this variability within a single hill, between hills, within plots, between plots, and in different years. Variability is a con-stant. Averages provide no more than a reference point and seldom reflect reality. There is virtually no such thing as a "normal" year.

During the study that my wife and I carried out from 2001 through 2003, we had no "average year" in south-central Min-nesota. All years were below average in rainfall, and that which we did have fell either fortuitously at just the right moment or fell at just the wrong time and contributed to viral and bacterial epi-demics in the beans and mold and bacterial spoilage of the corn. Heat was sometimes a problem because corn experiences heat stress that adversely affects yields when temperatures are above 86 degrees (30° C). Nevertheless, we were fortunate to have had very good yields because of precipitation at critical times. (See Munson 2004 and Munson Scullin and Scullin 2005 for data and sources.)

In 1988 and 1989, which were both years of extremely high temperatures and extreme drought in the Midwest, my demon-stration plot on the campus of the university at which I taught produced virtually no corn even though I watered it almost daily. The tassels "fired" (dried out) and produced little pollen because the plants could not pump enough water fast enough from the roots to the tassels. In contrast, 1993 was a very wet and cool year, with extensive flooding throughout the region. When I harvested in October (rather than in late August or early September as usual) the silk was still green on most ears and the corn was very moist and could not be stored until it was dried down to about the optimal 15 percent moisture.

We currently grow four varieties of Northern Flint/Flour corn from seed, which I obtained during the 1970s almost entirely from Hidatsa, Mandan, and Arikara gardeners on the Fort Berthold Reservation in west-central North Dakota. The women who provided me with seed back then were the last of the last generation to have practiced what some would call "traditional" horticulture. "Traditional" is, in my experience, an ever-moving target. My conclusion is that if people recall their grandmother or grandmothers as having done something one way or another, that something is considered traditional. If "these seeds" were planted by a person's grandmother, then the seeds are traditional.

By the 1970s the constantly evolving "original tradition" was already almost two centuries in the past. By the beginning of the nineteenth century, iron tools—iron hoes in particular—had become the norm. Buffalobird-woman never used anything but an iron hoe. The only bison scapula hoe that she knew was the one owned by her grandmother, who kept it under her bed. The kids were told to keep their hands off.

New seeds were being introduced by the time Lewis and Clark were passing through in the first decade of the nineteenth century. New plants and varieties continue to be introduced.

Using watermelons (*Citrullus lanatus*) as an example, documentation of them being grown in the area dates back more than two centuries (Clark 1804). By the time anyone asked the right questions and bothered to write anything down, the gardeners insisted that they had "always" had them. But it is now well documented and accepted that watermelons originated in Africa and were introduced to the southeast coast of this country in the late sixteenth century, from whence they rapidly made their way into Indian communities.

Melvin Gilmore, one of the first and best known ethnobotanists of the Northern Plains, devoted quite a bit of time, effort, and scholarship to demonstrating the indigenous origin of the watermelon. He concluded, "The watermelons grown by the various tribes seem to be of a variety distinct from any of the many known varieties of European introduction" (1919: 69). Alphonse de Condole, though, had already concluded by 1886 that "at length it was found indigenous in tropical Africa, on both sides of the equator, which settles the question" (1886: 263). Leonard Blake documents the introduction and acceptance of watermelons in what is now the United States in detail (1981: 193–99). If anything defines watermelon, it is the term "sweet," and sweet was notably in short supply in the diets of most Indians, though much desired. It did not take long for the sweet and seedy watermelon to make its way from the Atlantic coast to North Dakota.

The same can be said for squash. The original local squash was probably a variety of *Cucurbita pepo* (Blake and Cutler 2001: 131), which (in our personal cooking experience) has little taste and likely functioned as "something else" to put in the soup. When various and far more tasty Hubbard squashes (*C. maxima*) were introduced, I think they quickly replaced the indigenous variety so that within a very short period of time they became the preferred (and within a generation or so the traditional) varieties. It seems to be a *C. maxima* variety that Buffalobird-woman was growing, and a widely printed picture of Sioux Woman (Buffalobird-woman's daughter-in-law) hoeing a row of squash in the garden with a bison scapula hoe (Wilson 1917: 14) is a picture of Sioux Woman pretending to hoe *C. maxima* to demonstrate the bison scapula hoe that Wilson had had made for the American Museum of Natural History.

All of the varieties of squash that I collected at Fort Berthold were described as being "traditional," but all turned out to be

C. maxima. By the time I visited the Fort Berthold Reservation in the mid- to late 1970s, some introduced plants could well be said to have been grown by grandmother and hence had seemingly been around forever. I have, in fact, heard, "This is the way Grandma used to do it" from almost all the gardeners with whom I spoke. The only references to archaeological sources for *C. maxima* and *C. moschata* in North Dakota are to be found in historic sites—Fort Berthold (Like-a-Fishhook) and Rock Village (Blake and Cutler 2001: 131).

Gardeners, of necessity, have to be pragmatists. True enough, there are strong cultural factors influencing choices and tastes. But something that tastes really good is likely to be given a try. The same might be said for a plant that is obviously superior or at least fits in nicely with a preexisting preference. If the plant produces a sweet fruit, then adoption is even more likely—corn and watermelons would be very good examples. The Hidatsas would eat almost a third of their corn crop as "sweet corn" (although, while Northern Flint is not technically a sweet corn, it is very sweet when picked in what gardeners and farmers call the green or "milk" stage). A Green Corn Ceremony was celebrated in horticultural villages throughout the eastern midcontinent.

By the time villages were being established in eastern North America and in the Great Plains about a thousand years ago, corn as well as some squash/pumpkins and sunflowers (and very possibly some *Chenopodium berlandieri*) were grown. Beans were added a bit later (Smith 2007).

I favor the hypothesis that corn made its way to North Dakota from the Southwest, where a similar variety of maize known as *maiz de ocho* (corn with eight rows of kernels) has been grown for centuries (Brown and Anderson 1947; Galinat and Gunnerson 1963; Upham, MacNeish, Galinat, and Stevenson 1987; Doebley 2004.). Because the Mandans and the Hidatsas came from the east, there is undoubtedly some infusion of eastern genes inasmuch as they probably brought seeds with them. Maize adapted to hot and dry weather is simply going to do better in the summers of central North Dakota than maize adapted to the generally cooler and moister eastern environments.

Transmission from one group to another of both the corn itself and the details of its cultivation, preparation, storage, and consumption can occur very quickly if the taste and yield are good. I do not for a moment doubt that the sweet green stage was one of the major factors in its acceptance. Furthermore, maize is what is called a very "plastic" plant, which means that it is capable of making rapid adaptations to varying environments. Selection of seed is always based on what grows well in a given year—gardeners are going to pick seeds from the best-looking ears or best-producing plants for planting the following season. Given the variability of each year's climate, the plant, in this case maize, that does best under most conditions will rather quickly be achieved.

The Northern Flints appear to have been derived from southwestern varieties but have achieved enough genetic distinction that some investigators feel it would be justified to classify them as a separate species (Doebley et al., 1986). According to Doebley (2004, personal communication):

> Strictly speaking [the term] Northern Flint should be reserved for the native flint corns of the northeastern United States. However, the corn of the Mandan and other Plains Indians is very closely related, so some have referred to Mandan corn as Northern Flint. It would probably be best to call the corn "Great Plains Flint" or "Great Plains Flour" and mention its relationship to Northern Flint.

Although Great Plains Flint and Flour may be the preferred name, Northern Flint seems to have widest acceptance and so is used here.

Flint corn has a starch (endosperm) that is harder than the starch of flour corn, and the kernels have a pearly luster. Flour corn has a softer starch, larger kernels, and is easier to process. The Hidatsas, Mandans, and Arikaras used wooden mortars and pestles to pulverize the corn to a cornmeal rather than using grindstones, which can be used to produce a true corn flour. Virtually all Indian horticulturalists growing maize at its northern limits used wooden mortars and pestles. Flint corn, in our experience, is more resistant to mold and bacterial spoilage, and the productivity of flint versus flour tends to vary slightly with the nature of the growing season. Northern Flint comes in a great variety of colors and color combinations, but maize color is contained in the seed coat and, as Buffalobird-woman notes, it all tastes the same.

In our test plots, yields have ranged from 20 bushels per acre to almost 40 bushels per acre (1 bushel of shelled corn = 56 lbs. or about 25.4 kg) depending on rainfall and the distribution of that rainfall during the growing season. As previously noted, when it rains is often more important than how much it rains. The area in which the Hidatsas live receives an average of 15 inches of rain per year, most of which falls during the growing season. Summers are extreme, and winters are extreme. Northern Flint handles a great deal of stress successfully and will ordinarily mature in fewer than 100 days, but that number too depends on the weather.

The town of Williston, North Dakota, is on the western edge of the territory discussed in this book, and the distribution of rainfall there is much like that of the reservation. The critical months of June and July average about 2.5 inches of precipitation each (about 5 inches, or almost 13 cm, with an annual average of 14.37 inches or 36.5 cm) (National Weather Service 2011). This makes maize cultivation possible but always a gamble.

Late spring frosts are not devastating until after the fourth leaf emerges because the growing point of the corn plant is underground until then. Early frosts, unless they occur before mid-August, are not such a problem because the corn can be picked and salvaged, sometimes with little or no loss of yield. In thirty years of growing corn I have never had a loss due to frost; when I lived in south-central Minnesota (twenty-seven years of gardening) I tended to plant in late May and sometimes early June. In 2010 and 2011 all the corn in our plot (now in south-central Iowa) was picked and dried before the end of August.

Northern Flint is short (about 5 to 6 feet tall, or 1.5 to 1.8 m) and often bushy with several stalks (tillers)—most of which may produce some sort of an ear even if it is only 4 inches (10 cm) long. Ears usually have eight rows, although some have ten. Tillering, incidentally, is an adaptation that increases the amount of photosynthesizing surface on a plant and thereby creates more energy for the plant to mature more rapidly (Scullin 1989). Whether a plant tillers or not seems to be a complex interaction between the plant's genome, weather, and planting density. Some plants tiller, and others do not. Ears are borne low on the plant, some of which appear to come from the soil line or some below it if the corn is hilled.

Northern Flint has to be "hilled," which means that earth is hoed up around a cluster of five to seven corn plants planted in a circle about 10 to 12 inches (25 to 30 cm) in diameter. Hilling is done as the corn begins to tassel or when it is about 18 inches high (45 cm); hilling in a mound that is 6 to 8 inches high (15–20 cm) helps to reduce the tendency of this variety of corn to tip over in the high winds that frequently accompany thunderstorms. Furthermore, the corn plant produces special "prop roots" (adventitious roots) in the hill, which also contribute to the plant's stability. Should the corn be tipped over anyway (as happens almost every year, sometimes two or three times), when the soil has become firm again—usually the next day—the plant can be lifted carefully and the hill tamped with the foot so that the plant is upright. Each time the corn is tipped and restored to an upright position there is stress on the plants because fine roots are broken, and this usually contributes to a reduced yield.

After corn was picked it was thoroughly dried on a drying rack, which was constructed in front of each lodge, stored in large grass-lined pits, and used for both food and trade. For a full account, see Buffalobird-woman's description in *Buffalo Bird Woman's Garden* (Buffalo Bird Woman and Wilson 1987).

Beans (*Phaseolus vulgaris*) were grown as a supplement to the diet. They were not needed as a source of protein, as was the case with the Aztecs and other Mesoamericans, because there was seldom a shortage of protein given the abundance of bison. The beans did, however, add variety to the diet. We currently raise three varieties of beans from Fort Berthold: the red bean, the shield bean, and the yellow bean.

Buffalobird-woman says that the red bean was the only one she recalls from her youth, although she remembers it as having pink flowers (the currently grown variety has white flowers). The yellow bean thrives in hot, dry weather. I suspect that it is a late addition to the garden—probably historic and probably from the Southwest. The shield bean is large, and I suspect that this too is a recent introduction. Probably because of its large size, the shield bean requires a longer growing season than either the yellow bean or the red bean. Bean yield data are nonexistent.

Squash is a variety of pumpkin (*C. pepo*). The squash described by Buffalobird-woman was multicolored and, judging from photographs, was a bushy plant and not a vine (and probably *C. maxima*—Hubbard squash). Our personal experience with the squash/pumpkin called Omaha green and white striped pumpkin (a *C. pepo*) is that it has little flavor and, when dried, is tough and chewy with no discernible taste when cooked with other food. Buffalobird-woman noted that they often ate young squash in soup when the squash were very small and tender (at about 3–5 days). An Arikara family I knew prepared squash soup for grandmother with young patapan squash (a variety of *C. pepo*) when the squash was about 3 inches (7–8 cm) in diameter.

Once again it seems that variety at mealtime was the determining factor and that when squashes which actually did taste good (and, important, were sweet) were made available, they were rapidly incorporated into Hidatsa (and Mandan and Arikara) gardens and other predecessor plants were abandoned.

The question often arises about "the three sisters" or the supposed mutual benefits of growing corn, beans, and squash together. The three sisters story is the sort of story that people would like to believe. Supposedly the corn provides support for the beans, which climb the stalks. The beans in turn, through the *Rhizobium* bacteria that live in nodules on their root systems, provide nitrogen to the corn; these bacteria can convert atmospheric nitrogen into a form that can be used by the bean plant and ultimately the corn as well. So the corn supposedly gets a boost for helping the beans. Finally, the squash, with their big leaves and sprawling habit, cover the ground, reduce weed competition, and reduce water loss on sunny days. Thus a garden is a big happy family of sisters helping one another out.

But nature does not usually work that way. Organisms tend to be competitive. The notable exceptions seem always to be highlighted when they are discovered, and therefore many such stories of noncompetition or cooperation (mutualism or symbiosis) have been published in the past few years. Still, organisms compete and overwhelmingly evolve mechanisms that enable them to cope with competition. And so it goes in the garden. The beans are really trying to steal sunlight from the corn, which is only a convenient way for the beans to get themselves off the ground and collect a greater share of the available sunlight. Every photon captured by a bean leaf is one less photon for the corn leaves. Therefore the corn does not share but instead has its photons appropriated by the beans. Photosynthesis is thus reduced for a corn plant that provides support for beans. Beans are also competing with the corn for water and nutrients.

The beans have evolved a mechanism that permits certain bacteria to live in nodules in the roots. The bacteria provide some useful nutrients to the beans (and vice versa) and thus provide justification for taking up bean space. The bacteria do convert some atmospheric nitrogen into a form that is useful to beans (and most other legumes), but there is not a grateful feed of nitrogen to the corn (Munson Scullin and Scullin 2005).

An unforeseen problem that we encountered was that in the three years that we ran our 100 ft × 100 ft (0.1 h) test garden (2001 to 2003), not once did the beans (planted as described in Wilson 1917) manage to climb up onto the corn. We tried tomato cages and had much better luck.

With the beans growing adjacent to the corn, rain beat down the beans and splashed bacteria and virus-laden soil on the bean leaves. The yield was very small compared with that of the beans growing on the tomato cages. We have continued to experiment, and in 2009 the beans planted immediately adjacent to the corn did poorly while beans planted with tomato cages providing support did well. I have on occasion had beans climb on the corn and thrive, but in our garden it is far from a sure thing.

Finally the squash (*C. pepo*) will take over the entire garden if given a chance, stealing all the sunlight that isn't intercepted by the occasional weeds growing where they can. Squash are hosts to both boring and sucking insects. They do best if carefully watched and the soil is mounded over the sprawling vines at points from which leaves and flowers emerge because new roots will grow from these points and take over from the main stem if it is destroyed by borers. In other words, they don't just grow and provide their fruits.

Buffalobird-woman never mentions any problems with any insects. We, on the other hand, are constantly looking for borers and squash bugs on the squash vines and beetles that eat the beans and the silk on ears of corn right down to the husk. With no silk there can be little pollination and few kernels are produced. Hornworms (caterpillars of a species of sphinx moth that grow to 4 or 5 inches, or about 100–125 mm) can denude an entire tobacco plant of leaves in a couple of days. And so it goes. It takes the intercession of the gardener or gardeners to manage everything and maintain a semblance of order. I've noted before that nature abhors a garden. Gardens quickly lapse into chaos without a lot of hard work. This truth Buffalobird-woman describes, and her strategy was to start very early in the morning and quit before noon when the full heat of a Great Plains summer can be overwhelming.

Gardens demand immense pulses of hard work at planting and harvesting times and smaller surges of hard work to do the hilling and keep the competing weeds from taking over. This was true even before the more aggressive European and Asian weeds were introduced inadvertently by soldiers, traders, missionaries, and reservation agents. Horses and food for the horses were the big vectors.

Gardening is a lot of hard work, but if things go well it can have a large and profitable payoff. Gathering of wild foods, on the other hand, does not have a large payoff except with some notable exceptions, like wild rice, which grows and was gathered in Wisconsin and Minnesota and could provide enough yield to support village-sized populations. (See Jenks 1901 for details of this system; Jenks was Wilson's mentor at the University of Minnesota.)

The goal of gathering, as far as the Hidatsas were concerned, was to gather both food that tastes good and materials (particularly wood) that could be used to create much of the infrastructure. Uncertainty is the curse of gardeners, especially in this region, which has little rain. Much of the ceremonial life was devoted to the production of a good crop, and this is the same the whole world over. But prayers were not always answered, and in some years people really had to struggle to survive.

Buffalobird-woman would have been twenty-four years old when three years of extreme drought began in 1863 (NOAA 2009), but she says nothing about it. The intensity of this three-year drought would have affected not only the gardens, which probably would have yielded nothing, but also the edible plants the Hidatsas gathered and probably the distribution of the bison.

When gardens performed at normal or almost normal levels, as in most years, then garden produce was a significant and welcome part of people's diets.

When it comes to plants the Hidatsas used to feed themselves, their gardens provided, by far, the most significant plants. The major problem facing them, as well as anyone else living in the Northern Great Plains, was (and still is) the extraordinary variability of the climate. Good years were always followed, at some point, by bad and really bad years. Winters could be bad, worse, or horrible. Summers could be hot and droughty or, like the summer of 2009, relatively cool and wet. In the end, the extremes of heat, cold, and drought set the limits on all plant life and therefore on human lives and cultures.

Gardens cannot move, but humans can. During the winter months, people resettled in smaller settlements down in the coulees or other sheltered areas to be near firewood and to avoid the wind. But during the worst hot and dry summers there could have been little to do but to seek the available resources wherever located. This probably meant following the herds of bison and living the lives of hunters in small, mobile groups. This strategy had been worked out more than ten thousand years ago and was well known and well established.

It was the gardens that made all the difference between the lifeways of the villagers and the far more mobile bison hunters like the Lakotas. Gardening compels stability, so the gardeners had to stay put to tend and protect their gardens or face a very good chance of losing their crops. Gardening forced humans to reorder and reorganize into larger and more complex societies. It also compelled many settlements to construct defensive palisades to help fend off marauding bison hunters and aggressive village raiders. Stealing is easier than gardening and could be immensely rewarding.

Gardening made most years more bearable during the worst times, as when the frozen landscape was at its least productive and mobility was limited by cold and snow. Having a ton or more of corn in a storage pit or pits was the sort of security that hunting and gathering could not provide. Gardens made it possible for fairly large groups of people (several hundred to a thousand or more) to live together (sometimes in the same location) for decades and even centuries.

Sunflowers (garden staple—oil and meal)

HIDATSA NAME: *mapi* (sunflower)
LOCAL ENGLISH NAME: sunflower
BOTANICAL NAME: *Helianthus annuus* L.

Sunflowers grow in great abundance in disturbed, sandy areas, particularly along roadsides, riverbanks, and "blowouts" where the topsoil and vegetation have been blown away, leaving a sandy basin. And whereas the question remains as to the exact area in which sunflowers were first domesticated, they were domesticated quite early—at least four thousand years ago, probably in eastern North America (Smith 2006: 1227). Early gardeners undoubtedly did what gardeners tend to do and saved the seeds of the biggest and best for planting the following year. Domestication was probably not a terribly long or involved process.

Sunflowers are hardy—the first seeds planted in the garden in the spring—and resistant to drought. The seeds are oily and therefore highly caloric, and they appeal to humans because of their nutty flavor. The fact that domesticated sunflowers and cultivated sunflowers are the same species and are insect pollinated means that there is always the potential for cross-pollination so that domestication remains an ongoing process of selecting seeds from plants with the most desirable traits.

Buffalobird-woman (vol. 18, 1915: 461–62)

Sunflower planting followed the breaking of the ice, as soon as the ground became soft and the snow melted. The name of the month was *Mapicce midic* (sunflower plant moon.)

Three seeds were planted in a hill, all in one spot. They were pressed down by a single motion, and all three held at the same time in the thumb and first two fingers.

Each family raised the kind of sunflower seed they preferred, some preferring one color, some another. But the varieties were fixed. Black seed grew black seeds, and white seed produced white. However we had little preference; all varieties had the same taste and smell. [Sunflowers are pollinated by insects and therefore impossible to maintain as pure strains when planted in the vicinity of other variations.]

White sunflower seeds when pounded in the corn mortar turned dark. I think it was the parching that caused this.

(On questioning, Maxidiwiac [Buffalobird-woman] insisted that large sunflower heads measured 11 inches across [28 cm], not counting the fringe of yellow petals surrounding the head. She measured with an old hat. Goodbird affirms that this is no exaggeration, as he knows from his own observation. One of these big heads yielded a double handful of seed or even a little more than that. GLW)

The large sunflower heads yielded seeds that were less oily than the seeds of the smaller heads. When seeds of the latter were pounded and the meal was taken out and shaken in the closed hands, it soon became shiny with oil.

[Hidatsa sunflowers typically had one large head and a multitude of smaller heads at the ends of shoots which arise from the axils of the leaves. A domesticated sunflower might have a dozen or even more of these smaller flowers which are 4 to 5 inches in diameter and yield smaller seeds.]

For the winter our family put up two or three sacks of sunflower seeds each year. These sacks were about 14 inches high and 8 inches in diameter on an average [I estimate that would be about 10 quarts or 9.5 liters per sack]. No particular kind was used.

I have said that sunflower seeds that had been exposed to frost seemed oilier than seeds not so exposed. This effect of the frost was more apparent on seeds of the large heads than on the seeds of the small heads. The seeds of the large heads never had as much oil in them as the seeds of the small heads.

Sunflower seeds, parched and pounded into a meal, were often mixed with corn balls, to which it gave an agreeable smell as well as a pleasant taste.

[For further information see Wilson 1917: 16–21.]

[Sunflowers grown in our garden, in my experience, have to be hilled when they have reached a height of 2 or 3 feet (.6–1 m) rather than at planting. After many years of gardening, I concluded that Wilson (who was, as I have noted, not a gardener) did not fully understand the Hidatsa word for "hill," which, in the case of corn and sunflowers, probably means something on the order of "the place where we planted the seeds and subsequently hilled" or, as Buffalobird-woman said, "all in one spot."

Sunflowers, growing tall and having many "branches," blow over quite easily. Mounding soil about six to eight inches (15–20 cm) deep around the base of the growing plants stabilizes them considerably, allowing the sunflowers to grow additional roots into the hills. This greatly reduces but does not eliminate the chance of being blown over. This practice is exactly the same as it is for corn: not planting in a hill but hilling after the plant is about two feet tall.]

FAMILY *Asteraceae*–aster family
GENUS *Helianthus* L.–sunflower
SPECIES *Helianthus annuus* L.–annual sunflower, common sunflower, sunflower, wild sunflower

Corn-smut (edible)

HIDATSA NAME: *mapedi* (corn-smut)
LOCAL ENGLISH NAME: corn-smut
BOTANICAL NAME: *Ustilago maydis* (DC.) Corda

As unappetizing as corn-smut (a fungus) looks to most of us, it nevertheless tastes just about like the plant upon which it is growing. I have eaten both that which grows on the stalk and that which grows on the kernels. When ripe, it sometimes fills the ear with a truly disgusting black mass—same fungus, just different sites and appearances and stages of development. It took me ten years or so to work up the nerve to try it. It should be eaten while it is still green and firm. As it matures it turns black and slimy. When fully mature it turns into a dry and spongy puffball-like form. For a brief time it was served in upscale restaurants (using its Aztec name, *huitlacoche*) to unsuspecting customers who had never seen it growing. Although it sometimes seems to grow where there has been some damage to the corn plant, it also grows on (and in) kernels thoroughly protected by the husk. A growing season with more than usual precipitation seems to favor its growth, although the very rainy 2010 did not create an abundance of corn-smut.

Wherever corn travels, the fungus travels. Corn-smut was eaten in Mexico centuries ago, and still is. As corn was grown further and further north, the fungus accompanied it. I have no choice but to cultivate it each year in our Hidatsa garden.

Buffalobird-woman (vol. 38 [n.d., probably 1912]: 65 and A–65 and B)

Mapedi is a black mass that grows in the husk of an ear of corn. It is what you say whitemen call corn-smut fungus. Sometimes an ear of corn appears very plump or somewhat swelled, and when the husk is opened there is no corn inside, only *mapedi*, or smut. Or sometimes part of the ear will be found with a little grain at one end and *mapedi* at the other. These masses of *mapedi* or corn-smut that we found growing on the ear we gathered and dried for food.

There is another *mapedi* that grows on the stalk of the corn. It is not good to eat and was not gathered up at the harvest time. The *mapedi* that grows on the stalk is commonly found at a place where the stalk, by some accident, had been half broken.

We looked upon the *mapedi* that grew on the corn ear as a kind of corn, because it was borne on the cob. It was found on the ears of the grain that were growing solid or were about ready to be eaten as green corn. We did not find many *mapedi* masses in one garden.

Harvest and uses

We gathered the black masses and half boiled and dried them still on the cob. When well dried they were broken off the cob. These broken pieces we mixed with the dried half-boiled green corn and stored in the same sack.

Mapedi was cooked by boiling with the half-boiled dried corn. We did not eat *mapedi* fresh from the garden nor did we cook it separately. *Mapedi* boiled with the corn tasted good, neither sweet nor sour.

I still follow the custom of my tribe and gather *mapedi* each year at the corn harvest.

FAMILY *Ustilaginaceae*
GENUS *Ustilago* (Pers.) Roussel
SPECIES *Ustilago maydis* (DC.) Corda–corn-smut

Prairie turnips (tastes good, sweet)

HIDATSA NAME: *ahi*
LOCAL ENGLISH NAME: prairie turnip (Indian breadroot, frequently called prairie turnip or *tipsin*, the Lakota word for the plant)
BOTANICAL NAME: *Pediomelum esculentum* Pursh

Prairie turnips clearly fall into the Hidatsa category of what might be called "good things to eat that are sweet." As Buffalobird-woman said of boiled prairie turnips, "They are very sweet. That is the way I like them best." A good deal of time and effort went into the acquisition of sweet foods, and perhaps none more than went into the digging and preparation of prairie turnips. Prairie turnips are legumes with very specific habitat preferences to which Buffalobird-woman alludes. The soil should be well drained and the surrounding vegetation relatively low and somewhat sparse. The plant blooms in early summer, and by the end of July the part of the plant aboveground has dried out and broken off. Thus the stem with mature seeds can tumble across the prairie, spreading seeds as it blows. The root forms a tuberlike swelling about two inches below the surface in which the plant stores energy for the next year's growth, and it is this part that is pried from the ground with the digging-stick. Prairie turnips are still dug and prepared exactly as described by Buffalobird-woman. Braids of peeled prairie turnips are often sold at pow-wows and other events, and they are still considered to be treats.

The "bone-grease" mentioned here was a vital ingredient of plains village cooking. It was made by using a stone hammer to pound the long bones (such as a femur or humerus) on a stone anvil. When the bone fragments were about the size of a domino, the pieces were put into a pot and boiled to release the grease they contained. This was skimmed off and used in a multitude of ways in cooking. The remaining broth was the base for soup, and the bone fragments were discarded—by the millions. They are still there to this day, being exposed by gophers digging in the sites and archaeologists doing the same thing. Bone-grease was a labor-intensive product and valuable to the point that it was frequently used in exchange for goods or services.

Buffalobird-woman (vol. 20, 1916: 184–87)

Prairie turnips or *ahi* grow most thickly in a strip 10 miles wide along the Missouri on the north side of the river, and it is there that we find the biggest turnips.

[The "north side" of the river in the vicinity of Like-a-Fishhook caught the full brunt of the sun and was therefore drier and had sparse vegetation. This niche was most hospitable to prairie turnips.]

On this side of the Missouri—the south—we find turnips along the sides of hills but less abundantly. They are never found in the Missouri timber.

We dug them with iron hoes (mattocks) in my day. These tools were narrow and made purposely for this. But if the soil was stony, we used a wooden digging-stick because an iron implement would be spoiled on stony ground.

When we dug them out, the roots that had a smooth yellow rind we esteemed best. Such were sweet, and we were fond of peeling them of the rind and eating them raw.

A few women would go out to dig turnips with their husbands, usually a few miles from the village. They would camp on the digging grounds for the night.

A girl would often go in the company of a relative. Her sweetheart would follow to help her dig turnips and always took his gun along to give her protection, as when juneberries were picked.

Each woman in the group would fill a couple of saddlebags full of the turnips. These would be thrown over the saddle on the horse she rode.

When a family got back to the lodge from digging turnips,

the women friends would come in, and to each visiting woman would be given about ten turnips. These she would take and cut off the top and the tap root and peel off the root's thick rind. She peeled them with her teeth, peeling from the stalk end of the root downward. Then she ate the turnips raw, all but the heart [more fibrous core], and sometimes, if they were sweet, she ate the hearts also.

Turnip roots varied greatly. Some were soft, and some were hard. Some were sweet, and some not. Some were rotten inside or dead.

We prepared the turnips for drying by cutting them, slicing them, but the turnip rind was first peeled off by the cutter with her teeth. The slices would be vertical, and the heart was thrown away.

The slices were dried on a skin on the floor of the corn stage and dried in three or four days. If it rained, the turnips were covered and taken inside, but no harm followed even if they got wet.

When dry, they were put away in sacks.

Recipes

Put dried slices in corn pounder (wooden mortar and pestle) and pound to fine flour. Put this flour into a pot containing hot bone broth and add a little bone-grease. Stir until thick and eat with a spoon.

Or pound dried slices and remove from the mortar. Then put in dried juneberries and pound fine. The flour and pounded berries should then be pounded together in the corn mortar to mix them. Mix with hot bone broth and add a little bone-grease.

Or same recipe as above, except use chokecherries instead of juneberries. But [dried] chokecherries must first be soaked before pounding.

Sometimes the dried sliced turnips were boiled without pounding. After an hour's boiling, add dried meat to the boil. The broth poured off this was excellent.

Freshly dug turnips we often boiled whole with the rind on. These could be left in the water in which they had been boiled for two or three days and were still sweet. No meat was used in the boil.

We also roasted fresh turnips in the hot ashes, peeling off the bark when we ate them.

I remember once hearing of someone planting *ahi* seed in her garden, but they never grew, and the attempt was a failure.

Grizzly bears were fond of wild turnips. I never saw a bear eating turnips, but I have seen where the grizzly had been sitting and rinds of the turnip roots were lying on the ground there. Men who have seen them eating the turnips told me how they did it. The bear would sit up like a man and hold the root in his two paws and bite and pull off the rind with his teeth just as a human being would do. And he would eat the inside of the root only.

We thought bears dangerous. I used to carry a short gun, my own, when on horseback and going into the timber. I held my gun ready in case I met a grizzly bear.

Gilbert Wilson's summary and observations from interviews with Buffalobird-woman, James Baker's wife, Goodbird's wife, and others (vol. 8, 1909: 77–83).

Prairie turnips and the digging-stick

Prairie turnips were gathered in the month that juneberries are ripe. They were brought home in sacks. These sacks were round, and each held something less than a bushel They were wider at the bottom than at the top, and half a buffalo hide was used to make one sack.

Such sacks were used to store squash, dried meat, and corn as well as turnips. One or two of these sacks of dried turnips were laid in for the winter by each family. This custom is still followed; sacks of cloth take the place of the old-time kind.

The turnips were usually peeled. This was done with a knife or with the teeth. A strip of the peel or bark would be torn off with the canine teeth of one side of the jaw; the rest of the rind could then be torn off with the teeth or fingers or both.

Usually the turnips were cut into narrow pieces and dried. The pieces or strips were cut off vertically with the cut running with the grain. The narrow core on the inside was thrown away as it is woody and of no value. It took several days to dry the sliced turnips. They are very starchy, and if there is much moisture in the air, the pieces become sticky and quickly mold. When thoroughly dried, a heap of the sliced turnips presents a beautiful snowy appearance.

If the turnips were small, they were sometimes braided instead of being sliced. Buffalobird-woman says this was always done after first peeling them and that unpeeled turnips were never stored away. I have, however, purchased strings of braided turnips that were not peeled.

In old times the turnips were dug with the ash digging-stick. This was about three feet long and sharpened with fire. It had a slight bend in the lower and heavier extremity, giving it a little of the appearance of a spade with a pointed blade. The digging-stick was well rubbed with fat and roasted over the fire. This made it heavy and almost as solid as iron.

In using the digging-stick, the upper end was rested against the abdomen, which was protected by a folded robe or blanket. The digger raised herself on her toes and swung around in a half circle and back again. This drove the point of the digging instrument well into the ground.

The root of the plant is a single wiry filament that descends into the ground several inches and then expands into a tuber. It is this tuber that is edible.

Prairie turnips are eaten raw. Schoolchildren on the reservation take them to school much as white children do apples.

When sliced and dried, they are eaten uncooked, and the Indians seem rather fond of them. I have had a saucer full of them brought in and set beside me as I took down dictation—much as a whitewoman might bring in a saucer of crackers.

More usually turnips are cooked by boiling. The dried turnips require two hours to be thoroughly done but can be eaten with lesser cooking. Fresh turnips are boiled with the peel on. As they are handed around from the pot, they are easily stripped with the teeth or a knife.

If it can be obtained, a little bacon is thrown in the pot with the boiling turnips, if the latter are of the usual sliced-and-dried preparation. If the turnips are boiled fresh, the bacon is omitted.

A mess of boiled turnips, especially of the sliced-and-dried kind, do not fall ungenerously on a whiteman's palate. I usually bring back a package each summer from the reservation for my own use. A handful of the dried turnips will make a good meal as they increase in bulk in the boiling. The taste is not unlike our cultivated turnip, but more delicate and sweeter. There is more nutrition in a turnip than in a potato of the same size.

Fresh turnips are often baked. Most of the reservation Indians now have stoves and use them with considerable skill.

The following information was gotten at different times from Buffalobird-woman:

We cooked turnips by throwing them into boiling water; peeling them when about ready to eat. A fresh turnip unpeeled was cooked for about an hour.

Dried turnips were boiled longer. A peeled turnip, unless thoroughly dried, did not taste very good. Therefore we usually cooked fresh turnips with the peel on but dried turnips with the peel taken off. This was done before the turnips were put out in the sun to dry.

A fresh turnip boiled would keep sweet for a day or more.

Juneberries ripen at about the same time as the turnips. Often we gathered juneberries and boiled them. When pretty well boiled, we poured in a flour made of dried turnips pounded fine.

In wintertime we did the same, only the juneberries, having been previously dried, were then pounded also. The juneberries were boiled first, as they took a longer time to cook. The pounded turnips were then poured in.

Dried turnips were also boiled without pounding them up. They were cooked a long, long time.

Nowadays we often add whiteman's flour when we cook turnips.

At other times dried buffalo meat was put in with the turnips or buffalo fat.

The way I like turnips best is to boil them fresh, wait two days, and then strip off the peel and eat. They are very sweet. That is the way I like them best.

To bake turnips we buried the fresh ones in ashes with the peel on. We had no stoves then; we just buried them in the ashes and put hot coals over them.

Another dish we used to make was dried turnips pounded to a flour and mixed with hot marrow fat. (Buffalobird-woman always called this bone-grease.) Dried corn was also pounded fine in a mortar and boiled. These two dishes together made a very fine meal.

Buffalobird-woman (vol. 9, 1910: 110–13)

Prairie turnips in mythology: An excerpt from the "Story of the Grandson"

The Moon had married a Hidatsa woman, and they lived together in the sky. They had a son to whom the Moon was giving advice. Only a small fragment of this complex story is given here.

The Moon kept the Hidatsa woman for his wife. The Moon lived with his wife, and they had one child, a boy. He grew old enough to hunt birds. The Moon made some arrows and gave them to the boy. Moon said to the boy, "My son, do not shoot at the meadowlarks. If you do, they will cry out, 'Oh you one-eyed! Oh, your parents are bad.' What the Moon feared was that the meadowlark might call out something to make the boy remember the life on the earth.

(In Hidatsa-Mandan mythology, the meadowlark is always the talkative character and usually something of a gossip but not always a mischief maker. Very frequently the meadowlark brings a gossip message that proves of value to the recipient. Boys were taught that it was not right to kill the meadowlark. GLW)

The Moon also said to the boy, "You must not dig any woman (female) turnips. The prairie turnips which we dig on this reservation are of two kinds. They all have roots alike. These roots grow in bulbs. But the plants are different; some plants have no flower on them, and these plants we call the woman plant. The Moon feared that if the boy dug up one of these woman turnips, he might be able to look down in the hole in the ground to his mother's village.

(We may presume that the woman turnip had sympathy with the boy's mother, who also was a woman and through whom he was related to the earth people. However, Caucasian philosophy should be used sparingly in Indian subjects. GLW) [It should be noted that there are no male and female prairie turnips, just some plants that, for one reason or another, do not bloom in a given year.]

One day the boy went out hunting for birds. The meadowlarks are gentle and are not wild as are some birds. One of these came circling around the boy and singing, "Chu-chu-chu!" The boy was not forgetful of his father's warning, but the lark kept flying around and around until the boy became angry. He shot an arrow at the bird but missed. He shot again but missed. He shot the third time and missed. The bird now flew away but called out as it flew, "Why do you want to kill me? You who are away from your people? You who are away from your village?"

The boy began to weep when he heard these words. He returned home to the earthlodge and told his mother. She comforted him. "Yes," she said, "we belong beneath, on the Earth."

Again the boy went hunting and found a female turnip. He dug the turnip out of the ground. It grew in a coulee, and he had to dig deep, for it had a long root. When he looked down the hole which he had made and where the turnip had been, he saw a light. He came close to the hole and looked through and saw far below on this earth the five villages of the Hidatsa people at the mouth of the Knife River. The children were playing about the villages, and older men were playing the stick game. All were having a good time. The boy longed to be there.

FAMILY *Fabaceae*–pea family
GENUS *Pediomelum* Rydb.–breadroot, Indian breadroot, scurfpea
SPECIES *Pediomelum esculentum* (Pursh) Rydb.–prairie turnip, tipsin

Jerusalem artichokes (good to eat, sweet)

HIDATSA NAME: *kakca*
LOCAL ENGLISH NAME: wild potato (Jerusalem artichoke)
BOTANICAL NAME: *Helianthus tuberosus* L.

Jerusalem artichokes may grow in clumps or scattered about the landscape because the plant reproduces both by seeds and vegetatively through its tubers. As Buffalobird-woman observes, the tubers can vary from a light tan to reddish. They taste good but contain the complex sugar inulin, which is not digested by humans. In the lower intestine, the resident bacteria convert inulin into methane and carbon dioxide—thus the theme of the story about Itsikamhidish and his bout of extreme flatulence.

Buffalobird-woman (vol. 16, 1914: 264–66, and vol. 26: 147–51)

Our word for the plant that we translate as wild potato is *kakca* or *kaakca*. When we first saw whiteman's potatoes, we called them whiteman's *kakca*, because they looked so much like the wild roots.

Wild potatoes grow along the Missouri River down in the timber. We never planted them in our gardens ever, though I have already told you a story of how at first my people raised ground beans and potatoes (Jerusalem artichokes) in gardens near Devil's Lake until they got corn from the Mandans. But that is an old story that has come down to us. No one lives who ever saw wild potatoes grown in a garden.

I put wild potatoes in two classes. There is one kind of which the root is as long as from my fingertips to my wrist and looks yellow. Then there is another kind of which the root is smaller and looks yellowish or red. The larger kind has the sweeter roots. [Botanically there is only one "class," although some may grow larger on plants that fortuitously have been growing in a better soil or a soil which retains more moisture through the growing season.]

The stalks of the two kinds look just alike, and there is no other difference between them except that the smaller kind grows also on a smaller plant.

Both plants have blossoms just alike.

Down in the bottom lands along the Missouri there are places where wild potato plants grow very thick. And in the harvest season the ground would be dotted with marks where diggers had been at work.

Like a seed potato of the whiteman's kind, a wild potato root puts forth a sprout from an eye or little head on the root.

We gathered wild potato roots in the fall, when the ground became dry, and the harvest season lasted until the ground froze.

In the spring again, when the ground had thawed, we could gather wild potatoes as long as the plant was only an inch or two in height, but when the plant got about two feet high, the root would be all hollowed out; there would be nothing left of it but the skin. We now called the potato root "fly-away," meaning that the root had gone, as if it had flown away.

The fall harvest for wild potatoes, or the time to gather them, was just after the garden harvest was over, in the latter part of September. It lasted until the potatoes could no longer be dug in the frozen ground. Commonly the roots were gathered before we went to the winter village and afterwards as well.

The women and children usually did the gathering.

One way was just to pull up the potato plant; many of the roots would be pulled with it. But they were commonly dug up with a hoe, or a stick might be used. But the hoe was the recognized tool for gathering potatoes.

The digger collected the roots in a rawhide bag, which was made of deerskin, buffalo skin, or some other kind. The hair was scraped off, and it was oiled and pounded to make it flexible. The bag was made with a mouth that folded over on the outside. The scraped or (formerly) hair side was within. It was sewed with the edges turned inside.

The bag when filled with the roots was not especially heavy.

It was not our habit to get wild potatoes often. The women of a household commonly went out once or twice in the fall and perhaps once or twice in the spring.

The roots were eaten raw or they were boiled with fats. Boiling was for about a half an hour.

We also parched the roots like corn—that is, roots freshly dug. We parched them with fats, stirring them around with a stick.

We never dried wild potatoes, nor did any other tribe of which I am aware.

Wild potatoes were never stored for winter.

For cooking, wild potatoes were washed like ground beans in a native earthen pot or in a wooden bowl.

When eaten raw, the potatoes were not peeled. This was not necessary because the roots have no perceptible skin or rind; they merely have a brown outside.

Buffalobird-woman (vol. 16, 1914: 401–4)

Story of Itsikamahidish and the wild potato

[Itsikamahidish is one of the first beings and is
simultaneously benevolent and corrupt—in short,
a trickster, here also referred to as "Coyote Chief."]

Once Itsikamahidish was wandering around after a heavy rain. In a coulee in the woods he found a place where the rain had washed down the earth of the embankment on either side, washing out at the same time a number of wild potato roots. These roots were lying in the bottom of the coulee.

"Ha," thought the chief. Itsikamahidish was Coyote Chief. "I shall eat now and be full!" and he smiled for he was hungry.

He spoke to one of the roots. "Who are you, wild potato?"

"You already know me. Have you not just called me 'Potato'?" asked the root.

"Always every other chief has two names. What is your other name?" asked Itsikamahidish.

"I have no other. I am just Potato," the wild potato answered.

"Oh, yes, you have. Tell it to me!" said Itsikamahidish.

The potato got mad at this. "Yes, I have another name," he said sarcastically. "It is When-a-man-emits-gas-he-goes-up-high."

"Good," said Itsikamahidish. "That is a fine name," and he sat down and ate and ate until he filled up his stomach. Then he went away.

After a while he emitted gas, and he shook a little as he did so. "Good!" he laughed.

In a little time he emitted gas a second time. "Good, good!" he cried, but again he shook. "I must look fine. I must look fine," he cried. "I will go and call on my sweetheart. She will be proud of me."

He dressed and painted his face and even trimmed his hair a little.

Again he emitted gas, and this time he jumped a little. Still delighted, he ran around among his friends and showed them what strange things he could do. But now every time he emitted gas he jumped higher and higher. He got scared. He leaped to the height of the trees and then even higher. Now when he came down he was nearly killed by the fall.

He seized hold of trees and held on when he emitted gas, but the trees were torn out by the roots. He tried to retain the gas in his stomach, but could not.

He went about crying out, "Who can keep me from leaping when I emit gas?"

The broombrush, or buckbrush, cried out, "What is the matter with you, Itsikamahidish?"

Itsikamahidish took out his pipe and gave it to the broombrush to smoke. "I want to find what thing is the strongest in all the world," he said. "I want it to hold me. When I take hold of rocks or trees and emit gas, I tear them all out by the roots."

"All right," said the broombrush. "Take hold of me. I am strong."

"I want to emit gas now," said Itsikamahidish. "What shall I do?"

"Seize me with both hands," cried the broombrush. Itsikamahidish did so. When he emitted gas, his legs flew up in the air, but he was not precipitated upward himself. All day long he remained by the bush. Every time he gassed, his legs flew upward, but he saved himself from flying upward by holding on to the broombrush.

At the end of the day the Indians found him lying there. Itsikamahidish's legs and the lower part of his body were all swelled up, and he was a very sick man.

And that is the end of the story.

Thus you see that while potatoes cause one to emit gas, if you eat many of them, either raw or cooked, they will cause your body to swell up and you too emit gas, just as happened to Itsikamahidish.

FAMILY *Asteraceae*–aster family
GENUS *Helianthus* L.–sunflower
SPECIES *Helianthus tuberosus* L.–Jerusalem artichoke

Hogpeanut (good to eat)

HIDATSA NAME: *amaca ke* (bean-dug) or *awaca ke'e* (beans-digging)
LOCAL ENGLISH NAME: ground bean (American hogpeanut)
BOTANICAL NAME: *Amphicarpaea bracteata* (L.) Fernald var. *comosa*
(L.) Fernald

As is the case with so many of the plants gathered and eaten, it is obvious that the amount of time and energy expended to acquire and process them in no way approximates the return in calories. What often seems to be forgotten is the amount of time and effort people invest in things that they like as opposed to things that they need. Like prairie turnips, ground beans are tasty and add variety to a limited diet. Buffalobird-woman was a busy person but still willing to spend half a day gathering a quart of beans.

Buffalobird-woman, Goodbird, and Wolf Chief
(vol. 16, 1914: 268–72)

When the leaves had fallen off the trees in the woods we knew that the ground beans were ripe. Also we dug wild beans or ground beans in September a little while after the frost caught the vines. We also dug wild beans in the spring when they were just as good as in the fall. But when the vines in the spring had grown as long as a lead pencil the beans were not so good and did not cook so well.

I used to gather wild beans. I would take out a small cloth sack, something the shape of a flour sack but much smaller, and put the beans into it. Or I carried them home tied up in the corner of my basket. I did not use a packing strap for carrying home my beans, but I did carry the little tied up sack home on my back under my blanket.

In a half day, I could gather, unaided, about a quart of wild beans.

When we moved into our winter village we Indian women very

commonly went out to gather wild beans, and sometimes when we had not yet gone to the winter village I went into the nearby timbers at Like-a-Fishhook Village and got wild beans for they grew abundantly.

Wild beans grow only in the woods along the Missouri River. I do not think that they grow anywhere else but here or in the woods along rivers that empty into it. I never heard of any wild bean vines being found elsewhere.

I have heard that once the members of one of the men's societies—I do not know its name—disputed about which of the two foods was the best—catfish or wild beans. The society divided equally concerning the matter. Exactly one half of the members thinking that catfish tasted best and the other half saying that wild beans tasted the best. It was therefore decided by them that the two foods were equally esteemed.

We were fond of wild beans, and if we had the time we always tried to gather some in the fall and spring. We liked wild beans better than the cultivated kind.

I have heard that in old days wild beans were sometimes put away for winter use. A family would get a piece of log about two and a half feet long and hollow it out like a dish, or trough. The beans were put into this trough and covered over with earth. The trough was then put away inside the earth lodge, and the beans could be taken out and eaten at any time desired. Earth was put over the beans so that they would not dry out.

I have only heard of the custom, never having seen it done in my lifetime, nor have I ever known these wild beans to be kept for winter use in my lifetime. We always cooked them fresh, although we sometimes kept them over until the next day to cook, but never longer.

To prepare wild beans for cooking we washed them well, like potatoes. We washed them in our earthen pots or in wooden bowls, of which we had many.

We cooked wild beans by putting them into a kettle and boiling them, adding a piece of buffalo fat cut up small. Cooked in this way wild beans tasted good. Usually the wild beans were first put into the pot, and then the little pieces of fat were cut up and added immediately after. If the woman who was cooking the beans had not buffalo fat she put in a little grease instead.

At the winter village the people often hunted for mice nests. They took sharp sticks and tried the earth where they thought there were nests. Where they thrust the stick into the ground and it was soft and hollow they knew there was a nest there.

Once my husband and I found a mouse's cache. We were in a wagon driving through the woods when one of our horses stepped on a mouse's nest, and his foot broke through into the cache beneath. We dug out the beans that we found there. They amounted to about a half peck. [With eight quarts to a peck, half a peck of beans would be four quarts of beans or a little short of four liters.]

I never found a mouse's cache or beans by means of a sharp stick. Indeed I never found a mouse's cache at all except that one which my husband's horse stepped into. This nest was about a foot in diameter and about four inches thick.

I have also heard this story from old times; there was a woman—I do not know her name—who was very expert at finding mouse nests. In the autumn she would go out into the woods with other women who followed her. She would point to a place and say, "There's a nest. You go and get the beans." Then she would go around seeking carefully and would point to another place and say, "There's another nest."

It was said that she knew because where the nest was the ground was raised up into a little mound, and from this mound a little steam or smoke or vapor arose, so very slight that only this woman's eyes saw it.

Wilson, Wolf Chief, and Goodbird (vol. 16, 1914: 273–78)

I had gone to Wolf Chief's cabin and was sitting at the table waiting for him to come in when I improved the time by asking Goodbird to tell me something about ground beans. As he was talking, Wolf Chief came in, and the account was carried to conclusion by the two together. I have resolved the bits of information so that each narrator is credited with his own contributions. Both men are confessedly less well informed in the subject than Buffalobird-woman. In using the material presented by these two in this account the editor should make a clean distinction between what these two men saw and what they had heard. (GLW)

Goodbird: We call wild beans or ground beans as you say *amaca ke* (bean, digging vine), or dug bean vine.

We dug beans about the time corn is ripe, for the beans form on the vines about the same time as the kernels of corn form on the cob. Ground beans can be dug all fall, and in the spring until the middle of May.

Wolf Chief says that his father once found a mouse cache full of wild beans and that there was a woman in the village named Lone Woman who was skillful at finding these mouse caches of ground beans.

"My father," Wolf Chief says, "was skillful at tracking mice to their dens where they had their caches full of beans. If you want to find the nest," he told me, "watch and find a runway of tracks going in one direction. Then go a hundred yards or more in a circuit and find where other runways converge on the first. Go to that place where all these ways meet, and you will find a place raised up in a mound. Thrust a stick, sharp at the end, into the mound and you will find it full of beans."

Wolf Chief: Mice have the habit of collecting the beans in little piles preparatory to carrying them home to their nest. I have found these little piles—just a tiny handful—where they had been collected.

Sometimes the nest is made in a hollow log. (This should be verified. GLW)

In autumn, when the tribe went into the timber for their winter village, people would scatter out and hunt for these mice nests. Men, women, and children did this.

These mice nests were found down in among the willows, not in the big timber.

I think the mice that gathered these beans are what we call bear mice [voles—either *Microtus pennsylvanicus* or *M. ochrogaster*]. They are brown and have short tails.

The mice collected the beans, I am sure, in the fall. They have a little path, a trail, by which the beans are carried to the nest. In winter, when the ground is frozen hard, I feel sure the mice could not gather the beans.

I think it is not easy to find these mice trails in the timber in the fall when the leaves hide the ground.

My father, Small Ankle, I understand, used to track the mice in the fall before snow fell. The trails were little paths with no grass on them; and they grew wider as they neared the nest, made so by constant use by the mice.

I think the entrance to a mouse's cache was underground, but my father did not tell me, and I never saw but one cache, and that after it was opened, so I cannot be sure.

My father once took me to see a cache that had been opened. "There," he said, "is where there was a nest."

I looked and there was a hole in the ground about 18 inches deep and two or two and a half feet in diameter. It was the cache, opened. It had been covered, my father said, with soft earth.

The cache was lined within with stuff like cotton. It was fine dry grass and the bark of the weed from which we make Indian thread or cord—something like tow or hemp. This cotton-like lining was about four or five inches thick.

The beans in the cache were unmixed with any other kind of seed. At least, I inferred this from what my father said, but I may be wrong. The cache held about a bag-full, nearly a bushel I think—about as much I should say as that box there, (the box was 14 inches wide, 8 inches high, and 2 feet long. GLW) [The box described would have held about a gallon and a half or about 5.7 liters.]

In the cache were found eight mice that ran out when the cache was opened.

As I did not see the cache until it was opened, I do not know how tall the mound over it was.

This hunting of bean caches laid up by mice was a recognized custom with us in old times but is no longer followed by us.

I do not know how the mice carried the beans to the cache. I think they filled their mouths and carried them there. But I know that pocket gophers have a pocket on either side of the neck in which they carry their plunder. I and Charging Enemy found a nest of pocket gophers. It had wild potatoes in it, a few ground beans, and some kind of roots—probably of wild potatoes. It too was lined with grass, and a big gopher was in it. The cache or nest was raised up a foot or two above the level of the surrounding ground and was about three feet in diameter. Some roots were lying on the top and looked as if they had been gnawed. We thought at first that it was a mouse cache and dug into it.

Pocket gophers live both on the prairie and in the timber. They are fighters. Twice I have personally seen a gopher drive off a cat.

I own a cat that is a good mouser. One day we took it down to our well. I thought perhaps the cat would catch any weasels that might be around. I discovered a pocket gopher among some logs and punched it out with a stick. The cat made at it to seize it, but the gopher reared up, opened its mouth, and made a curious hissing noise and so frightened the cat that I could not get it to go near the gopher.

When I was small I used to hunt mice. If I wounded one and went close to it to attack it with a stick, the mouse would rear up on its hind legs, open its mouth as if to bite, and try to fight me. These are bear mice I am describing. I think the mice must be called thus from this habit. Bears also have the habit of raising themselves up in this manner, but I cannot say for sure.

Wild beans were boiled with a little fat. Sometimes we boys would get a handful and cover them up in the pot ashes to roast. Roasted or baked in this manner they were very good to eat, too. I cannot say whether the boys first washed them or not. I never cooked any this way that I remember.

FAMILY *Fabaceae*–pea family

GENUS *Amphicarpaea* Elliot ex Nutt.–hogpeanut

SPECIES *Amphicarpaea bracteata* (L.) Fernald–American hogpeanut

VARIETY *Amphicarpaea bracteata* (L.) Fernald var. *comosa* (L.) Fernald–American hogpeanut

Chokecherries (taste good)

HIDATSA NAME: *matshidumatu* (*matsu*, berry; *hidu*, bone;
matu', has or having = berry that has bones)
LOCAL ENGLISH NAME: chokecherry
BOTANICAL NAME: *Prunus virginiana* L.

Chokecherries are extraordinarily hardy and can grow to the
size of a small tree or be dwarfed to only a few inches (and still
produce a fruit or two). More often than not they are found in
clumps where usually a single plant has grown by sending up
shoots along its roots. Such clumps are often found where chokecherry eating birds have had a chance to perch long enough to
leave a seed or two behind. These clumps then become the nuclei
of "plant islands" in the prairie with grapes, raspberries, fragrant
sumac, and perhaps a juniper, forming their own self-sustaining
ecosystem. Occasional prairie fires keep these islands in check,
but they usually manage to recover. Chokecherries, like juneberries, are often found as edge (ecotone) plants between the prairie
and the gallery forests along the major waterways or on north-
facing slopes, but it is not unusual to find them growing as individuals in places where no other tree or shrub can survive (as on
the edges of rocky prominences).

Buffalobird-woman (vol. 20, 1916: 218–24)

Chokecherries are rather abundant and grow in the hills and in
the edge of the timber along the Missouri on the side next to the
prairie. The bushes are quite common.

We gathered the cherries when ripe—about the first of September. As soon as they were brought into the lodge, if anyone
wanted to eat them he got a handful and ate.

Or the fresh cherries were boiled with fats or bone-grease and
eaten as they were taken from the fire. Nothing else was mixed

with them. The eater spat out the seeds to be swept up later, or swallowed them—just spat the seeds out on the floor of the lodge, I mean.

But the chief way of preparing the cherries was to pound them up on a stone into a pulp, make into lumps, and dry.

The cherries were pounded with a stone hammer or with a hand stone.

In my day, the only man in the village who made stone hammers was Raise Heart.

[Stone hammers come in many sizes—from those so large one wonders who could heft them to the hammers Buffalobird-woman is describing which were standard kitchen tools. Some hammers were little more than cobbles with a groove pecked and ground around it for the attachment of a wooden handle with some green rawhide. Others were finished symmetrically and ground smooth. The stone was usually granitic. The kitchen hammer was about 2 feet long and weighed about 2 pounds. Larger hammers could be used for driving posts or breaking ice—sledge hammers. Large hammers were undoubtedly made on the spot where they were to be used and left behind when people moved on.]

Making stone hammers

Buffalobird-woman continues: Raise Heart was a magic man and made these stone hammers. *Maopaki* is the native name. We did not know how he made them. When someone wanted a stone hammer he found a suitable stone and took it to Raise Heart. "Come back for it at such a time," Raise Heart would say. If the man came back as directed he found the stone hammer all worked into shape. But if he put off coming and did not go to Raise Heart's lodge until after the time told him, he would find that when he came that Raise Heart had cut the stone in two!

Raise Heart had ghost power. I have heard that others have heard him working and that he made a noise like one striking a stone as he worked.

The stone hammer was a tool that a woman used. But very often a woman did not want to go to Raise Heart's lodge because he had this magic power, and she was afraid.

I do not know what he was paid for making a stone hammer. He had a large family. He was the only man in the village who did such work.

Very often instead of a stone hammer we used any convenient stone with a flat bottom and of a size and shape to fit the hand to pound chokecherries.

We used a larger stone for a mortar or anvil to pound the cherries on. In every lodge was kept such a stone all the time. Sometimes when not in use it was laid over the mouth of the corn mortar and kept the dust out of the mortar and protected it from dogs. This stone had three uses: we pounded chokecherries on it, we pounded dried meat on it to make it fine for pemmican and for eating—old people especially had to have dried meat pounded fine for them. Also we used the stone to break bones for boiling bone-grease. For this latter purpose we used a bigger hammer that the woman wielded with both hands. These larger hammer stones I do not think anyone ever made. We just found them either out on the prairie or elsewhere. And on them we put handles. Every lodge had one. Raise Heart made only cherry hammer-stones. The stone mortar or anvil we called *miimike*, from *mi*, stone.

Processing chokecherries

We pounded the chokecherries on the stone mortar in the lodge or out on the floor of the corn stage. We did not like to do so out on the ground outside the lodge on account of the dust that got in the cherries.

Two or three cherries were laid on the stone and struck smartly, then two or three more. When enough pulp had accumulated, it was taken up in the woman's hand, made into a ball, and then squeezed out in lumps through the first finger and the thumb of the right hand by pushing with the left thumb into the right palm.

These balls we dried on the corn stage on a skin. On warm days they dried in three days' time. But if the weather was damp and chilly, it might take five or six days. They were ready when a ball broke dry clear through. If one ball was put away while still soft inside, it spoiled and smelled bad.

Usually a family filled one sack with the dried chokecherry balls, some families two sacks.

The sacks were laid away for the fall in one of two places. Either on the food platform that stood in every earthlodge, or in the *makinudakci* or swinging loops that hung from one of the rafters back of one of the big supporting posts of the lodge. Every family had one of these swings.

For permanent storing for winter, the bag of dried cherries was put in a cache pit.

Now inside the lodge was a cache pit that was used chiefly in the season just following the harvest. The pit or pits outside the lodge were to hold the corn and other foods that we were to use when we came back in the spring from the winter village.

Now the sack of chokecherries was commonly put in the winter cache—that is the one inside the lodge. But if the family had an abundance, say, enough to fill a second sack, it was put outside in the next-summer cache pit.

Sometimes in the evening before going to bed my father would say to me or to my brothers, "Let us have some dried pounded chokecherries." Into the kettle I put warm water and soaked some of the dried lumps of chokecherries, leaving them for ten or fifteen minutes. Then I poured off the water and took the softened lumps to the corn pounder. I put two double handfuls into the pounder with enough raw fats to fill one's hand, sometimes first warming them in the fire. When well pounded together I took it out and put it into a bowl. Then I put in another double handful with fats, and sometimes even a third. The pounded results I made into egg shaped balls a little larger than a hen's egg—one ball to each person. We called this *matsupi* or pounded cherries.

2. The *makidudackci*, a mouse-proof hanging sling. Drawing by Goodbird.
(Courtesy of the Minnesota Historical Society)

Another recipe we called *mapiixti*.

I filled a dishpan half full of pounded parched corn. I put in three double handfuls of pounded, soaked, dried chokecherries without the fats. Into this I put bone broth—the broth left after skimming off the bone-grease. This I mixed together until thick enough to make into balls—three or four buffalo-horn spoonfuls of the bone broth were enough for this. Then I formed it into balls a little larger than dried chokecherry balls by squeezing in both hands. I now passed them around.

These lumps or balls could be carried a few days on a journey either in a heart skin or a paunch skin or in a skin bag if need be.

In old days we did not mix our foods as whitemen do—that is, we did not eat several kinds of foods at one meal. We ate when hungry, and quite often, and just one thing at a time, as we wanted.

Also when we made blood soup we mixed some softened dried chokecherry lumps with a little water and put them in the soup to give flavor.

If we had no chokecherries, we cut a green chokecherry stick and shaved down the bark on one end to form a ball of shavings and stirred the cooking soup so that it would smell of cherry flavor.

Dried chokecherry balls were a food that was valuable to correct one's bowels if they were too loose. It would stop the physicking [diarrhea] at once. But ripe chokecherries eaten raw acted in a contrary manner.

We Hidatsas were, and are, very fond of chokecherries. A family that ran out of them would go to the nearest neighbor and ask for some.

In old days we did not barter as we do now. We went to the next lodge of relatives or neighbors if needing food. Any kind of food needed was given or asked without thought of pay.

Gilbert L. Wilson (vol. 29, 1916: frame 0287)

Making chokecherry balls

August 16, 1916. Watched Owl Woman today
making chokecherry balls or lumps.

She used a stone hammer and stone mortar in olden style. The cherries were not quite dead ripe. She laid two to four cherries on her stone and struck them with her hammer, crushing them, seeds and all. Then three or four more cherries and so on until she had a handful of the crushed cherries collected, which she took into her palms and patted into a round ball. Letting this ball lie in her left hand and covering it with her right, she pressed down her right thumb and hand, squeezing the chokecherry mass out between the thumb and forefinger of her left hand at the same time drawing the hands backwards so as to deposit the cherries in an elongated lump on a board or slat she had provided for this purpose.

They were left on the slab to dry.

FAMILY *Rosaceae*–rose family
GENUS *Prunus* L.–plum
SPECIES *Prunus virginiana* L.–chokecherry

Buffaloberries (sweet—after being frozen)

HIDATSA NAME: *mahici* (red-thing)
LOCAL ENGLISH NAME: bullberry (buffaloberry)
BOTANICAL NAME: *Shepherdia argentea* (Pursh) Nutt.

Buffaloberries are a widely scattered resource but highly visible. The shrubs are 10 to 15 feet (3–5 m) tall with silvery gray foliage and often grow in clusters. The number of bright red berries on a single shrub would probably run into the thousands. They are members of the olive family, and accordingly they have formidable thorns so that picking is best done as described by Buffalobird-woman.

Buffalobird-woman (vol. 20, 1916: 251–52)

We gathered bullberries in the fall in parfleche bags and bore them home to eat raw. We would bend a tree down over a tentskin and whip it to knock down the berries. A man usually went out with his wife or perhaps with several women to gather them. Sometimes a boy went with his mother.

This was done after frost had fallen. Bullberries are sweet and good to eat after they have been frozen. The berries had no other use than to be eaten raw.

When the berries were brought home we winnowed them or put them in a bowl of water so that the little sticks and leaves and worms floated and could be readily removed.

Bullberries brought home to eat after frost were passed around to members of the family and to visitors. They were never dried nor saved to be eaten at a meal but just eaten as fruit at any time.

Bushes for fences

Bullberry bushes were used to strengthen tobacco field fences. [Buffaloberries have substantial thorns.]

3. Woman playing a game. Drawing by Goodbird. (Courtesy of the American Museum of Natural History)

Women's game sticks

Bullberry branches were used for making sticks for a kind of dice or women's gambling game.

Two sticks about a foot long were cut, carefully peeled of bark and smoothed, and then split. The pith was removed and the flat or split side also smoothed. Marks were made on the flat side. There were twelve chips or counters to go with them.

FAMILY *Elaeagnaceae*–oleaster family
GENUS *Shepherdia* Nutt.–buffaloberry
SPECIES *Shepherdia argentea* (Pursh) Nutt.–buffaloberry

Gooseberries (taste good)

HIDATSA NAME: *Mitektsatsa aku a apsa* (berry, kind of, fruit-plant, sharp)
LOCAL ENGLISH NAME: smooth wild currant (Missouri gooseberry)
BOTANICAL NAME: probably *Ribes missouriense* Nutt.

Buffalobird-woman (vol. 20, 1916: 260)

These berries taste almost like those of the smooth wild currant so that we give both berries nearly the same name. But these berries are a rather dark red when ripe, not black. Also before they ripen they bear white stripes on a green background. These white stripes disappear when the berry grows ripe. The berry is smooth, not prickly. It is very sour when green but tastes sweet when ripe.

This is a sketch by Goodbird drawn from memory, but the stripes should be white, not dark as in the sketch.

These berries ripen in juneberry time. Indeed we reckoned that all berries having small seeds ripened early in the season, while berries having one large seed each, like chokecherries and wild plums, ripened later.

We used to gather these berries and bring them to the lodge and distribute them, but we never dried them. Sometimes they were dropped in the basket when found by the women picking juneberries.

4. A currant. Drawing by Goodbird. (Courtesy of the American Museum of Natural History)

Like the smooth currant, this currant—for so we reckon it—was scarce. The bushes were not large, and there were not many of them. Bushes grow along the Missouri and in the hills.

FAMILY *Grossulariaceae*–currant family
GENUS *Ribes* L.–currant
SPECIES *Ribes missouriense* Nutt.–Missouri gooseberry

Black currants (sweet, face paint)

HIDATSA NAME: *Mitektsatsa aku a apsata* (berry, kind of, fruit-bearing plant, not sharp = berry of the kind of plant that is not prickly)
LOCAL ENGLISH NAME: smooth wild currant (American black currant)
BOTANICAL NAME: *Ribes americanum* P. Mill.

Currants or wild gooseberries are plants of the understories and edges of wooded areas and not usually found in full sunlight, They generally grow about thigh high and are among the first flowering plants of the year. The blooming of the currants coincided with the planting of the first corn.

Buffalobird-woman (vol. 20, 1916: 258–59)

These berries ripen when juneberries ripen in early July. Bushes were found along the Missouri and back in the hills. Berries were nearly or quite black when ripe. The bushes were not abundant, and the berries therefore were scarce.

The berries we thought very good eating, but they were gathered only casually and eaten right from the bushes. I do not think they were ever gathered to dry and store—they were too scarce for that. I never remember that they were gathered and fetched home to be eaten fresh.

They could be eaten freely, and one did not get sick from eating them.

They never dried by themselves. If the women of the village were picking juneberries and found some of these currant bushes, they often picked the currants and dropped them in with the juneberries. Then, of course, they were taken home, and they and the juneberries were dried together.

Young men gathered these currants when ripe and dried them for face paint. The dried berries they pounded with a stone and put them in a cup with a little water. Into this they stirred the pounded berries. The liquor so obtained they poured on some white clay. This was squeezed in the hands and made into a ball which was put away. It was a very good face paint.

When one wanted to use it, the ball was slightly wetted and rubbed on the face.

It made a pink color, while grapes made more of a purple color. Both colors were admired.

I have told you that grapes were squeezed fresh for the same purpose—painting the face. But I remember that my father used to keep dried wild grapes in his paint bag, I think he must have used them in the same way perhaps, but I can only guess at this. The bag he used to keep them in was a sort of general bag that we called *maaduxoki ici*, that is, "odds-and-ends sack" or "remnants sack"; *ici* means sack or bag.

FAMILY *Grossulariaceae*–currant family
GENUS *Ribes* L.–currant
SPECIES *Ribes americanum* P. Mill.–American black currant

Wild grapes ("good to eat," personal decoration)

HIDATSA NAME: *macipica* (black thing)
LOCAL ENGLISH NAME: wild grape (riverbank grape)
BOTANICAL NAME: *Vitis riparia* Michx.

Wolf Chief (vol. 20, 1916: 242)

Wild grapes were very good to eat when first ripe in the fall, and we Hidatsas ate lots of them. We ate them raw off the vines. This was our way of eating them. We never cooked them though we sometimes brought them home and ate them, say, the next day.

Another use made of them was to make red paint.

I took some white clay, powdered, in my palm. I took ten grapes and crushed them and pulped them in the clay with my fingers, mixing the pulp and clay in my palm. I picked the seeds out of the mass and threw them away. Then I added a very little water and touching my finger in the mass I tried a little on my wrist, letting it dry there. If too red, I added a little more clay. We wanted it to be a rose color.

When just right, I made a mass into a ball. This I used for painting my forehead and the hair over the forepart of my head and in front at the sides. Also the gum spots on a hair switch. The paint so put on the hair lasted one day.

[A rather elaborate adornment made of a young man's own carefully saved hair to make his hair seem a lot longer.]

The clay ball I put away. It lasted a long time and did not fade.

Added by Buffalobird-woman: Grapes were good to eat, but were eaten fresh off the vines. The paint was excellent and was much used.

FAMILY *Vitaceae*–grape family
GENUS *Vitis* L.–grape
SPECIES *Vitis riparia* Michx.–riverbank grape

TWO

Plants That Can Be Eaten

Hawthorns (food, but with qualifications)

HIDATSA NAME: *maamua* (I howl like a wolf)
LOCAL ENGLISH NAME: red haw (fireberry hawthorn or hawthorn)
BOTANICAL NAME: *Crataegus chrysocarpa* Ashe

In some prairies these small trees dot the landscape. Being very thorny, they are little browsed. The small, red fruits look appealing, but as Buffalobird-woman makes clear, they are nowhere near as desirable as they look.

Buffalobird-woman (vol. 20, 1916: 317–18)

Maamua grows in the hills and on the edge of the Missouri timber on the prairie side. It has red berries the last of September and first of October. The berries are good to eat but likely to give one the stomachache and swell up the abdomen. For this reason we feared to eat them very much.

When grizzly bears eat these berries too much they have stomachaches and their bellies burst and they die. Also, if one is scratched on the leg or arm by the thorns of this tree, the wound becomes sore and refuses to heal for a long time.

Before one of us ate of these berries he would howl like a wolf, "Hwu-u-u-u-u," and then would not have a stomachache. Every adult did this before eating. Perhaps children did not because they were careless. But we older people were rather afraid of these berries and did not seek them out very much.

Sometimes we gathered the berries and squeezed them fresh into our palms, forming a ball to take home. Sometimes we gathered them in a basket. We did not distribute them in the lodge but let them stand in a basket, and anyone who wished might help himself. They were not greatly prized as food.

We used to cook them. Three or four double handfuls were put in a pot and boiled with water and a little fat or bone-grease was stirred in with a stick. This stirring released the seeds which floated, for the most part, to the top and were skimmed off with a spoon.

We might have them like this once a season.

The name is from *mua*, or *wua*, howl (as of a wolf). The word means, "I howl like a wolf," referring to the custom I have described.

FAMILY *Rosaceae*–rose family
GENUS *Crataegus* L.–hawthorn
SPECIES *Crataegus chrysocarpa* Ashe–fireberry hawthorn

Wild white onions (children's treat)

HIDATSA NAME: *mika uti* (*mika*, grass, and *uti*, the foot of a tree or plant. That is, when a tree is cut down near to the ground, it is cut at the *uti*. I do not know why it is so called. Buffalobird-woman)
LOCAL ENGLISH NAME: wild onion (wild onion or prairie onion; also wild white onion (Stevens 1963: 103)
BOTANICAL NAME: *Allium textile* A. Nels. & J. F. Macbr.

Buffalobird-woman (vol. 20, 1916: 227–28)

This plant grows about six inches high. It has a white blossom and leaves and stalk like an onion. It has a bulbous root which tastes very much like an onion. We named whiteman's onions after them—*mika uti*.

The plant blossoms in May, about May 10—when the juneberry trees blossom. We thought it a good time when wild onions and juneberry trees blossomed for it meant that the frost was out of the ground and bean and squash planting could begin.

The plant grows on the prairie and is very abundant. We did not look for them in swampy ground, nor in the woods nor on the sandbars.

Children used to make a wooden pin or digging-stick and go out in early spring and dig onions up and eat them. They were the first plants to grow in the spring and children used to dig them up when the plants were an inch or two high.

I have heard that wild onions are a good medicine for one who has diseased bones. Those so afflicted gathered the roots and ate them often, and the roots went all through the blood and then through the flesh and bones and marrow. We knew this because animals that had eaten of wild onions tasted and smelled of them—a taste we did not like in the meat we ate. Even the marrow of the animal tasted of the wild onions. Hence we thought

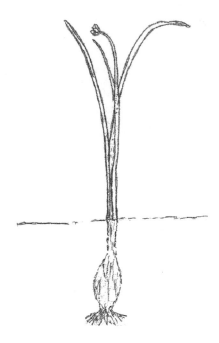

5. Wild white onion. Drawing by Goodbird. (Courtesy of the American Museum of Natural History)

wild onions a good medicine for diseased bones because we were sure the onion flavor went clear through the flesh and bones.

Older people did not like wild onions as children did, and the roots were never gathered to bring home. Only children ate them and then only in the spring—and raw. They were never cooked, either for food or medicine. Girls and boys both ate them. They made the children's breath smell.

Once, after we came to Independence, I gathered some of these wild onions and boiled them with flour. They were somewhat like rutabaga soup. I did this because I knew whiteman's onions were so cooked, but I never did it again because I did not like the taste.

FAMILY *Liliaceae*–lily family
GENUS *Allium L.*–onion, wild onion
SPECIES *Allium textile* A. Nels. & J. F. Macbr.–prairie onion, wild onion, wild white onion

Ball cactus (sweet fruits, emergency food)

HIDATSA NAME: *patskidia aku xaxua* (cactus, kind of, bunch)
LOCAL ENGLISH NAME: bunch cactus (spinystar—also commonly known as ball cactus)
BOTANICAL NAME: *Escobaria vivipara* var. *vivipara* (Nutt.) Buxbaum

"One girl would gather one heaping double handful in about half a day," Buffalobird-woman noted in 1916. The ball cactus (as it is known in North Dakota) generally grows scattered along well drained ridges and hillsides, but they can also be found growing nestled down in the prairie grasses, and the plants, although they often grow in clusters, are usually widely scattered. Thus it took half a day, a long time, to collect that double handful, even for someone who knew just where to look.

This is yet another "sweet plant" eagerly sought by people who had little sweet to snack on—this one with spines to cope with as well. They would serve it as an emergency food, largely to fill the void in one's stomach, because there are certainly very few calories in each plant.

Buffalobird-woman (vol. 20, 1916: 197–200)

There are three kinds of cactus growing on this reservation. These are *Patskidia aku cuka*, or wide cactus [*Opuntia polyacantha* Haw.–plains pricklypear]; *Patskidia aku tawoxi*, or small cactus [*Opuntia fragilis* (Nutt.) Haw.–brittle pricklypear]; and *Patskidia aku xaxua*, or bunch cactus or bulky cactus [*Escobaria vivipara* var. *vivipara* (Nutt.) Buxbaum–spinystar or ball cactus].

Wide cactus has a wide leaf that is flat. The spines on this cactus are larger than those on the bunch cactus.

The small cactus has no flower so far as I know. The wide cactus has a pink flower, but no berries, I believe. Neither the small nor the wide cactus were used by us. [Both the "small cactus" and the "wide cactus" actually have yellow flowers which are very attractive and produce fruits about the size of the tip of one's little finger.]

The bunch cactus was eaten, but only by warriors or hunters when they were out of food. Warriors dug the cactus up by the roots and laid them on the fire until the spines were burned off. The charred outside portions were then removed or cut off and the fleshy inside of the plant would be eaten. I never ate a bunch cactus myself. Only warriors when pressed by hunger ate them. They were cooked in only one way—laid on the coals or buried in them. I have heard that they taste something like green beans boiled in the pod and also resemble them in smell.

Bunch cactus bears a pink flower on the top. The plant bears from two to six berries, each a little smaller than a wild plum, and these get ripe just before frost or in September. They are good to eat and taste like figs. Whiteman's figs we call *patskidia ooki* or cactus head—ornament, in other words—cactus berries. I have wondered whether the whiteman's figs grow on big cactus plants or do they grow on trees? I do not know.

We gathered the cactus berries in small children's bark carrying-baskets and in parfleche bags. A parfleche berry bag was made envelope-shaped and from the chest part of a buffalo skin. The big parfleche bags that we made were from the back of a buffalo, and the berry bag was sewed on the sides, bottom folded up. A thong from either corner supported it on the berry-gatherer's breast, passing over her head.

[The large parfleche is a roughly square piece of leather in the center of which is placed dried meat or whatever one wishes to carry. It is folded in thirds and then the long ends folded inward, again in thirds. The package was then tied. Plains Indians often painted the exterior with geometric designs.]

Children and their mothers gathered the berries, especially girls. Two or three, sometimes four, girls went out together. We found the cacti growing on small ridges or rising ground. One girl would gather one heaping double handful in about half a day—counting a day in Indian fashion, for we did not work a whole day straight. Bunch cacti were not plentiful and were, therefore, hard to get.

When a girl came home to the lodge with her basket of berries everybody in the lodge, young and old, would cry: "Give me some! Give me some!" and each would hold out his hands laughing. The girl would reach in her bag and, drawing out the berries, would put some into each outstretched hand. The children especially would eagerly crowd around the girl, crying out and begging her for berries. The berries were not saved until meal time but were eaten up as soon as brought in. But then we had no regular meal time anyway. The berries were never cooked but were always eaten raw.

Girls over ten years of age gathered the berries. Girls of less than ten years would be likely to get spines in their fingers. Girls, after they were ten years of age, might pick the berries in season all the rest of their lives perhaps. Then too, everybody would pick and eat berries whenever they happened to find them in the prairie.

Boys would go out and gather cactus berries sometimes, eating as many as they wanted and carrying home the rest. They would take along a bag or basket for this purpose.

Ordinarily if the hand was wounded when one was picking berries, the wounding spines would pull out when the hand was withdrawn. Sometimes one stepped on a cactus, and the spines broke off in the foot. The spines were then pulled out with thumb and finger nails or with a pair of beard-tweezers—made of steel in my time.

We never tried to dry or put away cactus berries. Probably they could not be preserved, but of this I do not know. I only know that such was not our custom.

Cactus berries were apt to give the eater a slight physic, but no harm followed.

Gilbert Wilson and Wolf Chief (vol. 20, 1916: 201–2)

Eating cactus

As arranged, I went with Wolf Chief and hunted up some specimens of bunch cactus, the kind the Indians considered edible. Goodbird even found some with the fruits on them, but they were not ripe at this time of the year. The cactus grows in bunches or clusters made up of multiple bulbs or fleshy bodies about an inch or more in diameter [more likely to be about 2 inches (5 cm) or even larger] laden with spines. Wolf Chief separated one of these fleshy bodies from the cluster and laid it on the fire, rolling it about on the coals with the point of a stick. "We did not thrust the cactus directly on the stick, only rolled it about with the stick thus," he explained. When it was cooked, Wolf Chief cut off the charred spines and burnt rind with his knife. "Sometimes," he said, "we ate cactus raw; there were many of them eaten on war parties!"

Continuing, Wolf Chief said, "Once Scattered Village, His Mouth Is A House Hole (i.e., like a hole in a house) and I were out on a war party—just three of us. We saw no enemies, though we went about 80 miles. We set out to return. Now on this war party we had two guns, but we had little powder and feared to use much of it because we might need it if we met an enemy. We therefore had to eat cactus quite often. I remember that His Mouth Is A House Hole had a bow and arrows. We ate cactus, as I have said, to save our powder.

I forget how many days we were out. We had food with us when we started, but coming back we ate cactus as we got clear out of other food for three days and a half. We crossed a big trail of about fifty mounted enemies, I judge. Those three days we had only cactus to eat. Cactus was a kind of last-resort food.

This cactus could be eaten in early spring and all summer and fall. We also ate prairie turnips on war parties, but after they bloomed, the plants broke off close to the ground, and we could not find the roots."

Buffalobird-woman (vol. 26: 147, and vol. 16, 1914: 311)

Cactus eaten by members of a war party

I have never myself seen it eaten, but I have heard that cactus was eaten by members of a war party when they were hungry. I do not think it was eaten under any other circumstances. It was the little round kind that was thus eaten, for two or three kinds of cactus grow on this reservation. I have been told that a stick was thrust into the cactus, and it was held in the fire until all the spines were burned off. Then the outside skin was cut away and the heart or inside part was roasted on the end of the stick held over the fire. (*Escobaria vivipara* or bunch cactus. GLW)

FAMILY *Cactaceae*–cactus family
GENUS *Escobaria* Britt. & Rose–foxtail cactus
SPECIES *Escobaria vivipara* (Nutt.) Buxbaum–spinystar
VARIETY *Escobaria vivipara* var. *vivipara* (Nutt.) Buxbaum–spinystar, ball cactus, pink pincushion cactus

FAMILY *Cactaceae*–cactus family
GENUS *Opuntia* P. Mill.–cactus spp., prickly pear species, pricklypear, cholla
SPECIES *Opuntia fragilis* (Nutt.) Haw.–brittle cactus, brittle pricklypear, fragile cactus, jumping cactus

FAMILY *Cactaceae*–cactus family
GENUS *Opuntia* P. Mill.–pricklypear
SPECIES *Opuntia polyacantha* Haw.–plains pricklypear

THREE

Plants That Are Sweet

As one reads through this book it is apparent that sweetness is frequently mentioned. For example, one reason for the gathering of prairie turnips was because they are sweet, and Buffalobird-woman was quite emphatic about that. Cactus fruits are sweet, ripe currents are sweet, and so on. But it seems that the favorite sweet plant was the juneberry. Those growing in our front yard are the source of great competition between several bird species and the occupants of the house. Even the dog has quickly learned to pick juneberries. The expenditures of time and effort to acquire sweet foods do not fit subsistence models based on efficiency or maximization, but those seeking good things to eat do not worry about efficiency.

Juneberries (sweet)

HIDATSA NAME: *ma'tsu tapa'* (berry soft)
LOCAL ENGLISH NAME: juneberry (Saskatoon serviceberry)
BOTANICAL NAME: *Amelanchier alnifolia* (Nutt.) Nutt.
Ex M. Roem.

Juneberries are surprisingly adaptable shrubs/trees. I have found them growing along a west-facing bluff of the Missouri River where nearby neighbors were silver sage, ball cactus, and dropseed. They do best, however, where there is more moisture and so are usually found on north-facing slopes where the sun is not so direct and its drying action is muted. Here juneberries can be found with chokecherries and buffaloberries as well as the almost inevitable ash. Whereas it takes a seed to get a juneberry started, further expansion of the population can be largely or entirely by shoots coming up from the roots so that a patch may, in actuality, be a single plant—at least genetically. Ripe juneberries can be stripped off by the handful and are very pleasantly sweet. Chokecherries and buffaloberries are more astringent.

Buffalobird-woman (vol. 20, 1916: 210–17)

Juneberries are much used on this reservation even now, and the wood we thought very useful in old times. In fact we called the tree itself *mida'ka'ti* or real tree or genuine tree, because we thought the wood the best of the forest.

Juneberry wood is hard, and we used it in making wooden pins to stake out green hides to dry and for tent pins. It was the usual wood for making arrow shafts, being tough and unyielding. We used it for anything for which we required hard wood.

Juneberries were gathered when ripe to be dried and stored away for winter. The berries differed a good deal on the various trees. Some trees bore very sweet berries, others less sweet; and some bore large berries, some small berries. And besides these were white juneberries, of which I will speak later.

We gathered juneberries either by picking them off the tree by hand or we broke off a berry-laden branch and beat it with a stick, thus knocking the berries off upon a skin.

Picking juneberries was women's work, but the men helped break the branches. A husband would break branches and bear them to his wife, who took charge of them thereafter.

The picking season lasted about the month of July—the berries ripened the very last of June and lasted through the next month. In this time I used to go out picking juneberries about two or three times, but they were very plentiful some years and a single picking might be enough. We took no dogs with travois along. These we used for gathering fuel, and picking berries would take up all our time.

I used to go juneberrying with my husband or, before I was married, with my girl friends. Or a young woman might go with her love boy—sweetheart—who helped her. They worked together, but there were others working within sight of them. We did not all go in a big party, however.

We went early in the morning, while it was cool. We would gather, each woman, one or two calf-skin bags full.

It was, as I have said, a common thing for a young man to help his sweetheart pick juneberries. A young man might send word to his sweetheart by some female relative of his own saying, "That young man says that when you want to go for juneberries he wants to go along with you!" Or else he watched when she came out of the lodge to start berrying—for it was not our custom for a young man to talk openly with a young woman.

Sometimes a young man who went out thus to help his sweetheart talked with her as they worked. But very often they worked in silence. Young people who did talk with each other always had marriage in mind.

A young man who sent word to his sweetheart would not wait

and walk with her to the berry-grounds but would go on ahead and wait in the woods for her to come. A young man who would go to the girl's lodge and talk openly with her we thought was something like a fool, and if he did such a thing the girl would get mad at him.

When a party of pickers went out together, the men usually walked together by themselves.

Always a man carried a gun or some weapon with him. Enemies might be lurking in the woods—indeed that is why the men went. Also there were grizzlies about, and they were dangerous. For these reasons the women were glad to have the men along and always safer if they were.

The pickers ate a breakfast in the lodge before they went. They took no lunch with them, for they expected to return soon after noon.

When a couple was working, either a man and his wife or a young woman and her sweetheart, the man kept his gun lying at his feet as he worked.

Of course a young man gave all his attention to his sweetheart—at least as much as possible.

Once I was out picking juneberries with Pink Blossom. A young man named Old Bear wanted to get after Pink Blossom, but she refused to encourage him even though he persisted. Old Bear made a big pocket in his robe under his arm—as Indians do by folding the robe—and filled it full of berries he picked and emptied them in Pink Blossom's sack. But she was angry and threw them out. She did not want him to pay her attention.

The pickers went home a little after noon. When it was time to go, the women bore off the heavy sacks, never the men; but first the women put sweetleaf [*Agastache*–hyssop] plants or ash branches or other leafy branches over their backs so that the juice that leaked out of the sacks might not soil their garments. The mouths of the sacks were tied shut, and the sacks were carried like firewood—packed on the back.

If the woman was accompanied by her husband, he went home with her, walking before her Indian fashion. Or, if they wanted to talk, he walked at her side, either on right or left. But a young man did not walk home with his sweetheart. As soon as the berries were picked, the young man went off by another road. He did not follow her or watch her with his eyes. I don't think that the young woman even bothered to thank him.

Now I will tell you how I prepared the berries

As soon as I got home with my berries, I ate dinner. That over, I laid many buffalo-hides or an old tent-skin on the corn stage and spread the berries there to dry, The berries were often winnowed first by taking up double handfuls of them and letting them drop again on the tent-skin. We took them up in the two hands joined together. This cleared them of leaves and twigs. Or we winnowed with a wooden bowl because we tried to avoid using the hands as the berries were rather wet. This was done immediately after dinner. However, some families did not winnow until the berries had dried for a day. They generally used a wooden bowl for winnowing.

The berries were then dried in the sun for five days, sacked and stored away in one of the cache pits, either inside or outside the lodge

Also, while the berries were being dried, they were gathered up and put in the sack for the night, and put where it was necessary to hang things—just back of the left (facing door) rear main post. Here were two big loops of thong hanging from a rafter—the swing. In this anything could be swung quite out of reach of mice. To prevent mice climbing down the thongs, a circle of hide was cut and the thongs or ropes passed through them.

We very often took a supply of the dried berries to the winter camp with us.

Picking juneberries we thought had some danger in it. Once, before I was married, Rabbit Woman and Tattooed Forehead,

6. Mouse protection (a smaller mouse-proof sling). Drawing by Gilbert Wilson. (Courtesy of the Minnesota Historical Society)

two young women, were picking juneberries. Enemies found their tracks, followed them, and killed them both because there were no men with them to defend them.

Another time some people went out juneberrying, and a man named Skunk's Arm was along. He lay down and was singing, beating time on the stock of his gun. He heard a horse approaching but thought it one of his own party. It was an enemy, who shot Skunk's Arm through the fleshy part of his hip. Skunk's Arm sprang for the woods. The enemy feared to count coup on him [touch him to receive war honors] because Skunk's Arm had his gun with him.

Another time a party was out picking berries and enemies came up to fight the village. All the berry pickers hid, but the enemies caught a boy and killed him.

*These are ways in which we prepared juneberries
for eating juneberry balls*

We put water in a kettle and brought it to a boil. Took the kettle from the fire and threw in three double handfuls of dried berries. We then let it stand, but not on the fire. Next we emptied out the water, put the berries in the corn mortar, and pounded them—first one half, then the other. Of this we made balls or lumps by simply squeezing in one hand. These were passed around and were much liked because they were quite sweet, and in old times we had no sugar.

JUNEBERRY BALLS WITH POUNDED TURNIPS

Another way was thus: Put the berries in hot water, drain, and pound as above. Take dried prairie turnips and pound to flour in the corn mortar. Mix this flour with the pounded berries and a little bone-grease, stirring all together.

PUDDING WITH WHEAT FLOUR

We now use a new way, boiling the berries whole with sugar and wheat flour. Sometimes the berries are pounded instead of being put in whole. I like flour of wild turnips with the berries much better than with wheat flour.

MIXED WITH CORN BALLS

Sometime the berries, soaked and softened in hot water and pounded, were mixed in corn balls. Three double handfuls of corn were parched, pounded, and one double handful of pounded juneberries added. This then was mixed by hand. Then a piece of fat was put in the corn mortar and pounded. Of this little balls or lumps were made by squeezing with the hand. They kept unspoiled a long time.

EATEN RAW

When we were picking, the berries were often eaten raw from the trees. We were fond of them, but if one ate too many, they were apt to physic one. Therefore, when one had eaten many of the berries, he drank freely of water, which seemed to check the tendency to physic.

Wolf Chief (vol. 16, 1914: 111–12)

Wolf Chief's friend, Mouth Used as Mortar and Pestle for Berries

I was sixteen years old when I went on my first war party. The leaders were Tsecictac, or Wolf Eyes, and Imatsuipakcic, or

Cherry in Mouth as it has been translated by the whites on the reservation.

But Imatsuipakcic does not mean that exactly. In old days we had a custom of taking a buffalo horn, filling it with ripe juneberries, and crushing them to a pulp with a little stick as a pestle is used in a mortar. Boys and old people did this in the lodge or out in the woods where they found the juneberries. The pulp thus made was like a kind of jelly. The end of the stick was chewed up in the mouth so that it would collect the juice and pulp. It was then thrust into the pulp in the horn and the end of the stick carried to the mouth.

Now let us suppose a man, we will say, lay on his back and let someone use his open mouth for the hollow horn to mash juneberries in. The hollow log, or corn mortar in which corn is pounded, or the horn so used for juneberries or anything else used like a mortar, together with its pestle or pounder is called *ipakci*. *I* means mouth, and *ma-tsu* means berry or cherry or any small round fruit. The whole name means therefore "Mouth Used as Mortar and Pestle for Berries," but the agency whites have translated it as Cherry In Mouth.

My own name at this time was Kuahawic, or Coming. My name, Wolf Chief, was given to me afterwards.

FAMILY *Rosaceae*–rose family
GENUS *Amelanchier Medik.*–serviceberry
SPECIES *Amelanchier alnifolia* (Nutt.) Nutt. ex M. Roemer–juneberry, serviceberry

White juneberries (sweet)

HIDATSA NAME: *matsu ataki* (berry, white)
LOCAL ENGLISH NAME: white juneberry (juneberry, serviceberry)
BOTANICAL NAME: *Amelanchier alnifolia* (Nutt.) Nutt.
ex M. Roemer

White juneberries are merely aberrations, like white wild strawberries. They occur "in little groups" because juneberries propagate vegetatively with shoots as well as by seed. Seeds from white juneberries would probably produce plants bearing the usual dark red berries.

Buffalobird-woman (vol. 20, 1916: 264)

There are few white juneberry trees. They grow in little groups and are quite rare. They are found in only one place about Independence.

The berries of this tree are first green and then white when they are ripe, though they turn dark when they are dried. Real juneberries are dark red when ripe.

The wood of white juneberries has the same value exactly as real juneberry trees. Also the berries are used and cooked in exactly the same way. Berries of both trees are equally sweet. The trees of both varieties are, I think, of the same size.

Both kinds of trees grow on the same kind of land, but white juneberry trees grow in little groups or bunches.

FAMILY *Rosaceae*–rose family
GENUS *Amelanchier Medik.*–serviceberry
SPECIES *Amelanchier alnifolia* (Nutt.) Nutt. ex M. Roemer–
Saskatoon serviceberry

Wild plums (sweet)

HIDATSA NAME: *makata*
LOCAL ENGLISH NAME: wild plum
BOTANICAL NAME: *Prunus americana* Marsh

Plums, besides producing plum pits, reproduce by cloning, by sending up shoots from their root systems (just like juneberries). These shoots can often be quite a distance from the original plant. The result is plum thickets, which, like juneberries, are often all the same plant genetically. Thus plum thickets are a resource which, although scattered, can produce a lot of good tasting, sweet fruit within a concentrated area. Because they taste good, raccoons, foxes, and coyotes (and humans) can spread the seeds far and wide. Genetic variability within the larger population means that plums can ripen over both the months of August and September. Although Buffalobird-woman says October is still a month for picking ripe plums, that would be pretty late for that part of North Dakota. Because plums taste so good, the Hidatsas had a technique to hasten the availability of edible plums, as Buffalobird-woman explains.

Buffalobird-woman (vol. 20, 1916: 261–63)

Wild plums grow along the Missouri and in the timber in the hills. They are plentiful, and we found an abundance of trees in old times. We were fond of the fruit. The wood was of no particular use that I know of.

When plums were small, they were green, but as they grew to full size, they turned light colored but were not yet ripe. When they were thus light colored, we gathered them—say two or three gallons. Somewhere outside the village we dug a little pit or hole about a foot deep. The bottom and sides of the pit we lined with long sage bushes, the kind with seeds on them [*Artemisia ludovi-*

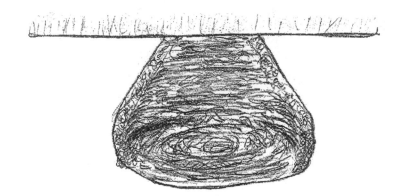

7. A sage-lined pit for ripening plums. Drawing by Goodbird. (Courtesy of the American Museum of Natural History)

ciana] two inches thick and very neatly. These white or light colored plums were then poured in, covered with two inches of sage and three inches of dirt.

Then the place was marked with a stick or bone or buffalo chip or grass. After four days we went out and dug up the plums and found them quite ripe. If left on the tree, the plums would not ripen until quite into autumn.

But when the plums in the little pit got ripe, the scent was quite strong, and the boys would smell the sweet scent and find the plums and dig them out and steal them.

The plums did smell a little of the sage that they were packed in, but not enough to hurt.

To keep the boys from stealing them, we had another way. We took a buffalo robe, turned it fur in, and wrapped the plums up in it into a bundle which we laid away in the lodge. And after four days—or four nights as we Hidatsas reckon—we found the plums ripe. We usually put the package or bundle on the *makidudackci*, or swinging loops, where things were hung to be safe from mice and dogs. Either method of ripening the plums was equally good.

Plums could be thus treated in August. On the trees they could be eaten only in September and October.

When they had ripened on the tree, we used to pick a sack-full and bring them to the lodge where anyone could go and help himself. Such a sack-full in our family would last about a week—in quantity about a bushel. However, when dead ripe, the plums turned black and sour after about four days.

I have heard that the Sioux and other tribes gathered wild plums and dried them, but we Hidatsas never did. We raised corn and vegetables, but other tribes had to gather wild fruits, not having our garden foods. Having plenty of foods ourselves, we did not have to gather inferior wild foods to store away.

When a sack of ripe plums was brought home and laid by for anyone to help himself, it was usually put on the platform where we kept foods—meat platform we called it. It was about as high as my daughter's sewing machine. There was but one in a lodge—but it might be higher or lower as the owner might desire. Each one chose the height that was thought convenient when building it.

Fresh plums were eaten boiled and were very good. They should be picked about this time—August 18th—while still green, but the seeds are now hard. Boiled plums were not sweet, nor yet sour—rather like the canned cherries we now buy at the trader's store. When boiled a long time, green plums turned a light color.

Boiled plums were not a frequent dish. But when they were made they were usually made by children, girls of 10–13, and

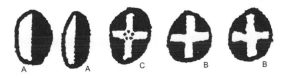

8. Hidatsa gaming pieces made from plum pits. *A* and *A* are edgewise and plain on one side and black on the other. *B* and *B* are marked with a cross on one side and plain on the other. *C* is like *B* except that it has a small mark on the center of the plain side, here marked with dots because it is on the reverse. Drawing by Goodbird. (Courtesy of the Minnesota Historical Society)

9. Mandan gaming pieces made from plum pits. All are edgewise. Drawing by Goodbird. (Courtesy of the Minnesota Historical Society)

always in the lodge. The dish was not thought of as an adult's food and was eaten without sugar.

Plum seeds were used for the game we called *atsuuke*. (No meaning that I know to this word, for it is not a compound, I think.) Five seeds were used.

The Mandans had a similar game, but the markings are different.

FAMILY *Rosaceae*–rose family
GENUS *Prunus* L.–plum
SPECIES *Prunus americana* Marsh.–American plum

Strawberries (sweet)

HIDATSA NAME: *amaaxoka aku apadumi* (ground-kidney,
of the sort that is short) or *amaaxoka aku ahatski*
(ground-kidney, of the sort that is longer)
LOCAL ENGLISH NAME: strawberry
BOTANICAL NAME for the "longer": *Fragaria vesca*
L.–woodland strawberry
BOTANICAL NAME for the "shorter": *Fragaria virginiana*
Duchesne–wild strawberry

Buffalobird-woman (vol. 20, 1916: 343)

These plants grow at the edge of the Missouri River timber where it skirts the prairie or hills. Their color is not dark red but a pure red. The plant is not high. Berries taste like the *amaaxoka aku ahatski*. The berries are very soft with no seeds in them and are small. They are so soft you can press them against the roof of the mouth with the tongue and crush them. The berries of the *amaaxoka aku ahatski,* the longer kind in the hills, are firmer.

The plants have a kind of hair over them, and the leaves are of color and shape something like rose leaves but somewhat larger.

Goodbird's wife interrupting:

These berries, if left standing after they are picked, get very soft in a little while.

Note by collector:

This berry seems to be the wild strawberry of the prairie, growing along the edge of the timber that lines the Missouri. I could get no specimens. GLW

FAMILY *Rosaceae*–rose family
GENUS *Fragaria L.*–strawberry

SPECIES *Fragaria vesca* L.–woodland strawberry

FAMILY *Rosaceae*–rose family
GENUS *Fragaria* L.–strawberry
SPECIES *Fragaria virginiana* Duchesne–wild strawberry

Roses (good to eat, make tea, smoke)

HIDATSA NAME: *mitskapa a* (rose bush). *Mitskapa* properly
means the rose pod [rose hip], Goodbird says. *A* means
a fruit-bearing bush or tree or plant. GLW
LOCAL ENGLISH NAME: rose bush–wild rose,
prairie and woods varieties
BOTANICAL NAMES: *Rosa arkansana* Porter–prairie rose
Rosa woodsii Lindl.–Woods' rose

Buffalobird-woman (vol. 20, 1916: 233–39)

Rose bushes are found on the prairie and in the woods. Those
on the prairie are short and rather scattered, and although they
have large berries, they are harder to gather because the bushes
are smaller and less abundant. In the woods the bushes grow in
groups or patches and are often as much as 5 feet high. The pods
or berries are smaller, but as the bushes are thicker, they are easier to gather. We called both bushes by the same name.

Rose pods we thought pretty good food. They were sweet like
sugar, and the pods of both varieties taste just the same. The pods
were gathered by women, generally when they were out after
wood. When the pods were gathered, as each one was plucked
from the bush, the woman picked the tiny leaves off the top of
the fruit so that only the pod proper remained.

The pods ripened in the fall. A good time to pick them was
after the first snow.

For picking them a small skin bucket or basket made from the
scrotum of a buffalo bull was used or a small bark basket. The
scrotum basket was used very commonly by all berry pickers.

Rose pods that I had gathered I put into a corn mortar as soon
as I got home and pounded them up, seeds and all. The pounded
pulp I took out and, having melted a little bone-grease, poured

10. A scrotum basket. Drawing by Goodbird. (Courtesy of the Minnesota Historical Society)

it in with the pulp and stirred it up. It was ready to eat at once. The mixture was sweet, and it had a pleasant smell.

I have been told that the Crow Indians and perhaps other tribes dried rose pods to put away for the winter. They mixed the scrapings of hides with the pods. They, however, used only the inner scrapings of a dry hide, those that come off after the outer dirty part is all scraped off. These clean scrapings were mixed with the rose pods, and all pounded up and dried.

They would lay a rawhide down, put the rose berries on it, mix them with the hide scrapings, and mash and pound them all together with a stone. Then this was dried, and when it was wanted for food, it was boiled. We Hidatsas never dried rose pods, however, to store them for winter.

Nor did we Hidatsas have frequent servings of rose pods. A family had them just once or twice in the fall or winter, as a rule. Once a Crow Indian came to visit my tribe. He was married to a Hidatsa woman, and they came back to visit her tribe. His name was Apaducic or Ear Eat. His wife's name was Yellow Blossom.

Ear Eat went into the woods by the Missouri and was crazy to see them full of rose bushes hanging thick with rose pods. He got off his horse, took off his leggings, tied them at the bottoms, and filled them with rose pods. He came back and called to his wife, "Here, look what I have—aren't they good?"

His wife was so shamed that she got mad, for being a Hidatsa woman she did not care much for rose pods. She took one of the leggings and poured out the pods violently on the ground. Then she did likewise with the other legging.

The Crow Indian of course expected his wife would dry the pods. In the Crow country the Indians set great store by them, and pods are scarcer there.

We Hidatsas also made tea from rose bushes. One would cut off a bush about two feet from the ground. Then the woman trimmed off all the smaller branches and shaved the bush clean of thorns and the outer red bark. Next the woman shaved off the inner bark and made a ball of it 3 or 4 inches in diameter, which was bound together by a strip of the same bark.

A pot of water was brought to a boil. Into this boiling water the woman put the ball of bark. In a little while she lifted this out with a stick. Into the tea thus made she added sugar, and it was ready to drink. This custom also came in only after we got whiteman's sugar.

Also, when men could not get kinnikinnick, they dried rose leaves over a fire and smoked them with tobacco. A man held a bush over the fire, and then, holding the bush over a skin or something, crumpled the ends with his hands, and the dried leaves crumbled and fell off. This was often done by members of a war party who were out of tobacco. I don't think it was a very good substitute, however.

Pipestems were made of rose bush stems, for they were hollow. Also rose bush stems were used to make rabbit snares to put in the woods. A stick of rose stalk so used was about a foot long, and it was split. I have seen these snares made, but this was not woman's work, and I have forgotten just how the fastenings to the stick were made. Older men made these snares; perhaps Wolf Chief can tell you.

For a pipestem one of the larger stalks of the rose bush was cut, trimmed of thorns, and split. Then, with a knife, the man scraped out the pith making the stem hollow. The split stem he put together, binding it in about four or five places with sinews. No glue was used, but the smoke never leaked out.

In the winter we used rose bushes about the winter lodges. I have told you that we built little earthlodges down in the timber to winter in. Now the dogs were likely to be troublesome, getting up on the roof of the lodge on account of the warmth there, and they would make holes in the roof by lying there and digging with their paws. Thus the family leaned rose bushes against the lodge thickly—stems to the ground and branches upright. The thorny bushes kept the dogs from climbing on the roof.

Bullberry [buffaloberry] bushes were used for the same purpose, but rose bushes were best. Sometimes the two kinds of bushes were intermixed and both used. The bushes were leaned quite around the lodge, clear up to the porch entrance.

Sometimes when a young man went to see his sweetheart, he would put a rose berry in his mouth, and when he kissed his sweetheart, his breath smelled sweet. This, I have heard, was an old time custom. I never actually saw it done myself in my time, but I have often heard that it was done in former times.

Flavoring broth of liquor amnii with rose berries

My father was a good buffalo hunter. Now it is August, and next month will be September. September is the autumn month in which we begin to drink calf-waters. These were very good until December when the buffalo calf got larger and the taste of the water changed. (Calf waters being the amniotic fluid. GLW)

We never wasted this calf-water. We thought bone broth—the

broth that is left after skimming off the bone-grease after boiling bones—was good, but calf-water broth was better. We boiled meat in it and then drank the broth that was left. We did this with calf-waters of the female buffalo, deer, antelope, and elk.

When my father had killed a buffalo cow in the fall, he butchered the animal so that the uterus was taken out unbroken. One of the women took the uterus, holding it by the mouth over a kettle, and made a slit in the side of it with a knife. Through this slit the water ran out into the kettle, and the calf itself fell out. The woman took the calf from the kettle into which it had fallen. She then added one-fourth water and set the kettle to boil, putting in meat enough to be just covered by the water. When about half boiled, the woman added the carcass of the unborn calf also, having first removed the entrails.

One-fourth water was added to the calf-water because unless this was done it would burn in cooking.

The calf-water was done when boiled for fifteen or twenty minutes, but one might wait a little longer on the meat that was added. This might be dried meat or fresh meat, either one. The calf was added when the meat was half done. The calf in September was quite small, like a gopher. It did not take long to cook the calf carcass.

This calf-water broth tasted good and sweet. Now adding salt to the broth makes it sweet and good—now that we use whiteman's salt. But this calf-water broth tasted sweet and good without adding any salt—just as good as salted soup that we make from other kinds of meat.

This calf broth besides tasting fine itself made the meat boiled in it taste good also. But if one boiled meat in this calf-water a second time, it made the meat taste too strong.

Now rose pods, or berries, were ripe in the fall—in the season when we got calf-waters from buffalo cows. Very often we flavored the broth by putting rose berries in the boil. My father once told me that it was an old custom of the tribe to put seven

rose berries in the kettle. But I always used two rose berries—
seven berries made it taste too much like roses.

FAMILY *Rosaceae*–rose family
GENUS *Rosa* L.–rose
SPECIES *Rosa arkansana* Porter–prairie rose

FAMILY *Rosaceae*–rose family
GENUS *Rosa* L.–rose
SPECIES *Rosa woodsii* Lindl.–Woods' rose

Red raspberries (sweet)

HIDATSA NAME: *amaaxoka aku ahatski* (I understand this word
to be from *ama*, ground or earth; *xoka*, kidney; *aku*, of the sort;
and *ahatski*, long. BBW)

LOCAL ENGLISH NAME: (none given) (common red raspberry)

BOTANICAL NAME: *Rubus idaeus* L.

Buffalobird-woman (vol. 20, 1916: 326)

(I feel sure this must be a raspberry. GLW)

This bush or shrub grows in the hills about the height of a woman's chest. I do not think I ever saw the blossom. The berry is rather long, like a kidney. I think whitemen put these berries up in tin cans. The color of the berry is red.

We gathered the berries when ripe—eating them off the bushes. We never tried to dry them. Sometimes we gathered up a pint or a quart and passed them around to eat. We gathered in rawhide berry baskets.

The berry grows on a kind of bush with leaves much like those of the sweetleaf plant [*Agastache*–hyssop]. It is not a branching plant—one single stem comes up and bears the fruit.

Goodbird's wife (Sioux Woman) interrupting:

These bushes have thorns on them, but they are not very sharp.

Buffalobird-woman resuming:

I never saw these berries except in berry time. They taste and look like strawberries (raspberries?) in a can taste. I do not know whether the plant is a woody plant or not.

Note by collector:

Buffalobird-woman means that she lived in the village and did not get out to the hills where the berries grow nor looked for them except when they were ripe. She describes the berry as long. Goodbird thinks the plant dies down each year and is not woody—but he is in doubt. Mrs. Goodbird, however, explains by saying that the bushes grow up in new shoots each year—which describes the raspberry's habit very well. She says they are ripe when just a few juneberries are ripe. GLW

FAMILY *Rosaceae*–rose family
GENUS *Rubus* L.–blackberry
SPECIES *Rubus idaeus* L.–common red raspberry

Biscuitroot (sweet)

HIDATSA NAME: *peditska ita ahi pi ca* (raven's turnip digger like)
(Goodbird was in some doubt of the meaning of this compound.
Apparently the translation is "like a raven's turnip digger." The root
of the plant is long and round like a digging-stick such as the Hidatsa
women use in digging wild turnips. GLW)
LOCAL ENGLISH NAME: wild carrot (biscuitroot)
BOTANICAL NAME: *Lomatium foeniculaceum* (Nutt.) Coult. & Rose

Buffalobird-woman (vol. 20, 1916: 229–32)

Wild carrots were dug up by girls, say, of about ten years of age.

Wherever there was some sort of ridge on the prairie, there the snows first melted, and it was there we dug wild carrots. We began digging when the plants were about a quarter of an inch high.

Digging time was about the latter part of April and lasted about twenty days.

We dug with a sharp stick held in the hand. This was usually a wooden pin used to stake hides out to dry because there were always many of these about the lodge.

Young girls when they dug the carrots peeled off the outer bark of the root (or rind as you say whitemen call it) and ate the root raw. They tasted fine, too! The roots were sweet and firm when dug early in the season, but when the plant had grown larger, the roots were no longer sweet.

I have often dug wild carrot roots and eaten them fresh, but I never cooked them freshly dug nor took them home to dry them. Nor did I ever hear of anyone in my tribe who took them home and cooked them fresh.

But my mother, Red Blossom, once told me that she knew a man's wife who gathered a lot of wild carrot roots and peeled off the rinds and mashed the roots with a stone and dried the mashed roots in the sun. When well dried, they were put away.

11. "Wild carrot," which is not a wild carrot but a biscuit root. Drawing by
Goodbird. (Courtesy of the American Museum of Natural History)

Later this man made a ceremony, and the feast given by him
at the time had for the guests a kind of soup made of the mashed
and dried wild carrots.

She said that the soup tasted and smelled something like wild
prairie turnips—*ahi* as we call them.

I forget the man's name who made the feasts. I never made
such a dish as this, or ever put away wild carrot roots in this
manner—or any other manner for that matter.

I suppose the mashed and pounded dried roots were put away
in parfleche bags or bags like those used for dried green corn,
but I never did this myself. I do not know, because the custom
of storing the dried roots had died out in my day.

Also once a number of Hidatsa people went to visit the Crow
Indians. Now the Crows are not corn raisers, and the Hidatsas greatly missed their maize foods to which they were accustomed. So they had their women go out and gather quantities
of these wild carrot roots and mash and dry them—first peeling off the rinds before mashing them. When cooked, these carrot roots made a dish that tasted somewhat like the maize dishes
the Hidatsas longed for.

In old times our tribe knew many plants and ate them each in its proper season, for they knew when each should be gathered. And because they knew each plant's proper season, no one got sick and no harm came of eating them. Such was our custom.

In old days, also, we had no whitemen with us and no guns, and we often faced scarcity of food. And so we ate such foods as wild carrots. But in my day it was not necessary to do this, and such wild foods fell into comparative disuse.

Wild bears—and by bears I always mean grizzlies—eat wild carrots. Once when Son Of A Star and Wolf Chief and Hairy Coat and Bear's Tail and their wives (eight persons in all were in the camp) were hunting on the Little Missouri River, they came upon a grizzly and two cubs digging wild carrots. The mother bear and the cubs both were digging them. The men did not molest the bears. The bears sat eating the roots like a man—sitting and holding their forepaws to their mouths.

The wild carrot plant has a yellow blossom. The leaves look much like those of whitemen's carrots. There are two or three blossoms on a plant.

Two varieties?

There are two kinds of wild carrots, one smaller and one larger. They are alike, except that one is a somewhat larger plant and has a larger blossom and stalk and root. The larger kind grows in places where there is clay.

We ate the roots of both kinds, and both were equally good.

The blossoms come in May. We never gathered the seeds, and I do not know what they look like. The plant dies down in mid-summer, and now (August) only the roots can be found in the ground.

(It is common on the prairie for the size of the plant to be determined largely by the kind of soil it grows in. I have found that Buffalobird-woman often divides a species of plant into a larger and smaller kind. She is a keen observer, and her statement that the two kinds are exactly alike probably means that there is but one variety. GLW)

[The other could well have been *Lomatium foeniculaceum* ssp. *foeniculaceum* (Nutt.) Coult. & Ros, also called biscuitroot or *Lomatium orientale* Coult. & Rose–northern Idaho biscuit-root, which is noted by Stevens (1963: 219) to be "common on prairie" and has white flowers in a smaller cluster.]

FAMILY *Apiaceae*–carrot family
GENUS *Lomatium Raf.*–biscuitroot, desert parsley, Indianroot
SPECIES *Lomatium foeniculaceum* (Nutt.) Coult. & Rose–biscuitroot, desert parsley, desert biscuitroot

Nannyberries (sweet)

HIDATSA NAME: *naxpitsi ita matsu* (grizzlybear his berry)
LOCAL ENGLISH NAME: bearberry, blackhaw (nannyberry)
BOTANICAL NAME: *Viburnum lentago* L.

Bearberries (the name is now usually associated with *Arcto-staphylos uva-ursi*) or nannyberries are fairly tall viburnums with smooth black fruits about the size of large raisins. Each fruit has a very bland flavor, a somewhat mealy texture, and a large seed. This probably explains Buffalobird-woman's lack of enthusiasm. The one characteristic that seemed to matter most was sweetness.

Buffalobird-woman (vol. 20, 1916: 305-6)

Bearberries grow along the Missouri on the side of the line of woods towards the hills, in places where hardwood flourishes. They never grow in places where cottonwoods flourish nor on sandbars.

When they are green and the berries just start to redden a little, we would gather the berries and wrap them up in a robe and lay them away to ripen as we did plums. But we never dug little caches for them. We gathered them about the first of September.

We generally ate them uncooked. Sometimes, when they were about half dried, we would boil the berries with some fats for a little while, and they would not spoil even after fifteen days. When we were cooking, we knew they were done when we saw that the water had turned all black.

Some nights about this time we ate some—about 8:30 PM—but not every night. We ate with spoons of buffalo horn.

When we gathered bearberries, men and women both gathered, one or maybe more together. We did not gather the berries in large quantities, and we never dried them to keep for winter.

The boiled berries tasted sweet, and when cooking they smelled something like peaches.

However, few persons cared to gather them, although there was an abundance of them in season. But like bullberries we did not especially prize them.

We ate the cooked berries as we ate other foods, at no regular time but just when we liked. Sometimes we ate in the evening and again at 8 or 9 o'clock.

Of our custom in eating, we ate, say, meat and broth in the morning. Then, in a little while, corn or boiled pounded corn with beans. We ate at all or any times, whenever one was hungry and called for food. We had no regular meals as whites do.

I do not know why the berries are called bears' berries. Remember, we Hidatsas reckon only grizzlies to be true bears.

FAMILY *Caprifoliaceae*–honeysuckle family
GENUS *Viburnum* L.–viburnum
SPECIES *Viburnum lentago* L.–nannyberry

Purple prairie clover (sweet)

HIDATSA NAME: *makadicta ita wika* (child his grass)
LOCAL ENGLISH NAME: children's grass (purple prairie clover)
BOTANICAL NAME: *Dalea purpurea* Vent.

Buffalobird-woman (vol. 20, 1916: 195)

The roots of this plant are long and have a sweet taste. Children chew the roots, swallowing the juice and throwing away the chewed cud. Young people also chew the roots. They are chewed any time in the growing months, summer and autumn. The roots were never fetched home and cooked. Children of three years of age on up to twenty years would chew them. The plant grows on sandy ground and to get the roots one just pulled it up.

I remember that we used to dig down around the plant for 3 or 4 inches into the earth and then pull out the root. We then peeled off the bark with our nails or, if we were at home, we scraped the bark off with a knife. This was just a children's custom, but sometimes parents or other adults would ask for some of the roots to chew.

The roots, I have said, were chewed at any time in the growing season, but they were sweeter in the fall.

The plant grows almost everywhere on the prairie, but we found it most easily dug when in sandy ground. It is not found in the woods along the Missouri.

The plant had no sacred use nor was it ever used as a medicine as far as I know.

FAMILY *Fabaceae*–pea family
GENUS *Dalea* L.–prairie clover
SPECIES *Dalea purpurea* Vent.–purple prairie clover, violet prairie clover

FOUR

Plants That
Are Good
to Chew

Sticky gum (chewing gum, adhesive)

HIDATSA NAME: *macika kadee* (gum sticky)
LOCAL ENGLISH NAME: sticky gum; probably from ponderosa pine
BOTANICAL NAME: *Pinus ponderosa* P. & C. Lawson

The northeastermost ponderosa pines are found in the badlands along the Little Missouri River in southwest North Dakota (near Medora) and in some southeastern counties in Montana (Barkley 1977: 12).

Buffalobird-woman (vol. 20, 1916: 347)

This is a gum found in a crack or broken place in the small branches of a pine tree in Montana. I remember my father climbing a tree and scraping off the gum with a knife. I spread a saddle skin under the tree and the pieces of gum fell on it.

The gum was brown as it came from the tree. On some smaller trees I have found some of this gum white and found it good to chew. "That gum was cooked by some prairie fire," said my father.

My father and I gathered a mass as big as one's two fists and took it home, and I boiled it. Using a stick thrust into the boiling mass, I tried a little with my teeth. When it did not stick to my teeth, I knew it was done.

The gum was carried in a heart-skin both before and after boiling. If boiled too long, the gum became like sand.

This gum we liked to chew, and it smelled good, too. Men and women both chewed it. Women chewing the gum would make it "crack" or "snap" with a sharp sound. It was liked because it was a breath perfume. Perhaps this is how the custom began. Chewing gum was our custom.

Besides its use for chewing, this gum was used in making switches—to fasten weasel skin straps to the tips of eagle feathers and the like. A bit was warmed and dropped on the feather and the strip made to adhere.

[Switches are used by males to make their hair look longer and fancier.]

Also some was put inside the wing bone of an eagle to make a whistle.

Wing bones of wild geese were also used to make whistles, that from a white goose [more probably a swan] being largest. The whistles of the Real Dog society members were made of white goose wing bones. The gum was used in them in the same way as in eagle wing whistles.

FAMILY *Pinaceae*–pine family
GENUS *Pinus* L.–pine
SPECIES *Pinus ponderosa* Lawson & C. Lawson–ponderosa pine
VARIETY *Pinus ponderosa* Lawson & C. Lawson var. *scopulorum* Engelm.–ponderosa pine

Pine pitch (perfume)

HIDATSA NAME: *macika iditsi saki* (gum smell good)
LOCAL ENGLISH NAME: good smelling gum; probably from
ponderosa pine
BOTANICAL NAME: *Pinus ponderosa* Lawson & C. Lawson var.
scopulorum Engelm.

Buffalobird-woman (vol. 20, 1916: 349)

Good Smelling Gum, as we called it, came from the same tree—a pine—but only in a certain place in the Rocky Mountains. Son Of A Star, when on a war party with some others, smelled this gum from quite aways away. It was on a certain hill that was called "Good-smell Hill." It was in the Assiniboine country. It was not known why the gum smelled so especially strong.

FAMILY *Pinaceae*–pine family
GENUS *Pinus* L.–pine
SPECIES *Pinus ponderosa* Lawson & C. Lawson–ponderosa pine
VARIETY *Pinus ponderosa* Lawson & C. Lawson var. *scopulorum* Engelm.–ponderosa pine

FIVE

Plants That Smell Good

Purple meadow-rue (perfume)

HIDATSA NAME: *mika' idi'tsi* (grass stink or scented grass)
LOCAL ENGLISH NAME: (none given) (purple meadow-rue)
BOTANICAL NAME: probably *Thalictrum dasycarpum* Fisch. & Avé-Lall. (Goodbird translates this term as stink grass. *Idi'tsi,* however, seems to mean simply "smell" and, as in English, if it was used without the qualifying word *sa'ki,* or "good," it usually meant "smell" in a bad sense. In this word it probably means "scented grass," and I should so translate it. GLW)

Buffalobird-woman (vol. 20, 1916: 206–7)

These plants produce seed that was much used for perfume. I gathered the plants about the middle of August. The seeds should be about ripe then. The seeds are brown when dead ripe and are at their best, but the plants should be gathered while the seeds were a little green though they were better when dead ripe.

Whether they were dead ripe or a bit green, I hung the plants up in the earthlodge in a bunch by the roots to thoroughly dry. I hung them on a post or some other convenient place in the lodge.

When the plants were quite dry I easily loosened the seeds by simply rubbing between my palms or fingers.

The threshed seeds we used as perfume, never, so far as I know, for incense or medicine.

One chewed up the seeds and rubbed them into his hair or on his clothing. The scent often clung to the clothing for two or three months. If I rubbed some of the seeds that I had chewed into my hair it lasted quite a long time.

In old days we Indians used perfume a great deal, and we therefore always smelled good.

If a young man walked ten or fifteen yards away scented with perfume, and the wind was right, one smelled him.

The seeds of *mika' idi'tsi* were often chewed and rubbed into the mane of a saddle horse.

However, when one chewed up seeds for perfume, he did not spit them into another's hair but used them always to perfume his own body.

Perfumes were in general use but were thought especially appropriate when people came together to have a big dance.

Stink grass seeds have no smell when dry until chewed up or wetted. We thought it a young women's perfume, though men and women both used it. The user chewed up the seeds and spat or blew them over the fur side of the wearer's robe or on cloth or rubbed them into his (or her) hair. However, some thought they bred lice when used in the hair, but I never found any proof of this tale.

FAMILY *Ranunculaceae*–buttercup family
GENUS *Thalictrum* L.–meadow-rue
SPECIES *Thalictrum dasycarpum* Fisch. & Avé-Lall.–purple meadow-rue

Blue giant hyssop (perfume, tea)

HIDATSA NAME: *a'pa tsiku'a* (leaf sweet)
LOCAL ENGLISH NAME: sweetleaf (blue giant hyssop)
BOTANICAL NAME: *Agastache foeniculum* (Pursh) Kuntze

Buffalobird-woman (vol. 20, 1916: 209)

This plant, sweetleaf, was used like horsemint, to be bound on fans for the sweet odor. As with horsemint, when the leaves got dry they were thrown away.

The name given the plant is from the sweet taste it has when one chews the leaves. When I was a young girl I often chewed the leaves and swallowed the juice. How general this chewing the leaves was in old times I do not know, but I remember that I and my girl friends all chewed them when I was young.

The plant was never used as a medicine as far as I know.

When I was about twenty [ca. 1859], sugar came into pretty general use in the tribe, and we began making tea of the leaves of the sweetleaf plant. I never heard of our doing this before we got whiteman's sugar. Sweetleaf tea is now quite commonly used on this reservation. I do not know who began its use.

We also began to drink tea made of elm bark about forty years ago after we got sugar, but elm bark does not make a very good tea.

FAMILY *Lamiaceae*–mint family
GENUS *Agastache Clayton* ex Gronov.–giant hyssop
SPECIES *Agastache foeniculum* (Pursh) Kuntze–blue giant hyssop

Sweetgrass (perfume, ritual)

HIDATSA NAME: *matsuatsa*
LOCAL ENGLISH NAME: sweetgrass (northern sweetgrass)
BOTANICAL NAME: *Hierochloe hirta* (Schrank) Borbás

Sweetgrass, as Buffalobird-woman notes, grows in damp soils (which are scarce where the Hidatsas live). It is still widely gathered by many groups and used both for incense in ceremonies and in some places for coiled baskets. It is a plant the Hidatsas would have put in a category of plants that can make a person smell good.

Buffalobird-woman (vol. 20, 1916: 188–92)

Matsuatsa grows in damp places—at the edge of a river or lake or pond, or in places in the woods along the Missouri where the floods overflow the ground. It is never found in the hills or on the prairie. Some is found on the Little Missouri near Spotted Weasel's. Some used to grow near Independence but has died out, I think.

The chief use of sweetgrass was for perfume. Young men especially kept it in their beds on the frame inside where they slept. Also, if a young man had a fan of eagle tail feathers, he would tie about four small braids of sweetgrass to the fan so that it smelled good every time the fan was waved. The braids were bound big end down, little end up so that they would flap every time the fan was waved.

There are seven feathers in an eagle's tail. The two middle feathers have the quill running right down the middle. In each feather to left and right of these two, the quill runs a little to one side of the feather. The feathers of the fan are similarly arranged, the left and right feathers of the tail going to left and right of the middle feather of the fan.

12. Sweetgrass braids on an eagle feather fan. Drawing by Goodbird. (Courtesy of the American Museum of Natural History)

On the extreme tips of these feathers go little strips of white weasel skin fastened with gum or glue, over which gum while yet moist is sprinkled some white powder made of burnt stone glass (mica, presumably. GLW) [This is more apt to be gypsum, which is fairly common in parts of the badlands just to the west.]

If a young man had a robe, he often tied sweetgrass through four or five of the drying pin holes [made when the green hide was stretched out on the ground to dry and held in place with wooden pegs] along the upper edge; these were not grass braids but loose grasses with ends tied together.

Braids of sweetgrass as I have said were hung from the inside of bed frames, especially those of young men. Braids hung from one of the poles of the frame that supported the covering of the bed or from one corner overhead. Such a braid smelled sweet for a long time. Girls and married people also used these braids.

Sweetgrass was used thus for perfume—just dried. Never was it chewed or used in any way other than dried the natural way.

In the old days young men and women were always scented up. Only the men, however, wore sweetgrass on the edge of the robe, as described. Nor did women use feather fans. Only men had fans. Old men used fans of eagle wings. Young men's fans were made of eagle tail feathers.

Sometimes a young woman tied sweetgrass on the ends of the braids that fell forward over the shoulders. The end of the braid was tied with a piece of cloth or thong. I used to make a little ball or bunch of sweetgrass and tied it with the same thong that tied the end of either of my hair braids.

This is a sketch by Goodbird of such a little bunch that I have made of common grass. Two average-length sweetgrass stalks were enough—even one sweetgrass stalk smelled strongly. The little bunch was about an inch long and was made by doubling the grass back and forth thus.

The sweetgrass bunch I took off at night and put on again the next morning. It was worn on one or both braids; usually the left if on but one braid.

Braiding the hair in two braids forward over the shoulders came into style in our tribe when I was a little girl. I think it came from the Dakotas and was adopted by us. Before that our women and old men tied the hair in a bunch over the forepart of the head with the rest of the hair hanging loose. They did not have to comb it then. Young men, however, braided the hair in two braids.

At night a young woman combed and braided her hair but left off the sweetgrass bunch, if she wore one. She braided and dressed her hair again in the morning after the morning meal. She used a porcupine tail for a brush, not the grass hairbrush of the Arikaras.

Older women wore a bunch of sweetgrass exactly like that I have described—tied to the collar or neck part of the dress on

13. Sweetgrass tie on Buffalobird-woman's braid. Drawing by Goodbird.
(Courtesy of the American Museum of Natural History)

the left side and fastened by a thong through an awl hole on the edge of the neck part (or what corresponds to the collar on a whitewoman's dress).

I even wore a ring of sweetgrass in the same position. But this was a ceremonial use. I did this in the Wolf Ceremony. A woman whose husband bought this Wolf Ceremony received and wore as part of the rite this sweetgrass ring. It was a braid of sweetgrass formed into a ring.

Sweetgrass had importance in use in ceremonies and medicines and sacred rites. In these it was used in many ways. But I am unwilling to tell these things because I am not a Christian.

It was used as incense to burn before sacred objects, but this again was determined by the rites of the particular ceremony that was observed. [Sweetgrass is still important to many people and used in many ceremonies.]

Sweetgrass could be gathered at any time in the summer season and smelled good.

[This is an annotation from a list of items purchased by Wilson for the American Museum of Natural History.]

Braid of Sweetgrass

(15) of Mrs. Foolish Woman. Her own name is Maksipka, a Mandan word, and explained as a small rodent animal, mouse colored, living in the ground, has a short tail. Size described about that of a prairie gopher or a ground squirrel of the eastern states. Apparently a mole is meant, but I am not sure.

She is forty years old, and claims to be a doctor. But this is a very usual claim on this reservation.

Braid of sweetgrass. Is duplicate of one in her bag of native medicines which I could not buy. Latter came down to her from her grandmother. Some red and yellow paint has been rubbed over it. A pinch off the end of this ancient and not very clean braid is used thus: "When someone has a spasm and they come for me to cure the patient I take a little piece off the end (about a half inch in length) and lay it on a few coals and hold in the sweet-smelling smoke the medicines (herbs and roots) that I am going to use on the patient."

Goodbird described how his child was cured of the spasms. "She went back to her house and in about twenty minutes returned with the medicine. It was steeped in water, and it looked blackish. 'There,' she said. 'Give the child one teaspoonful. If he sweats well, it will be enough. Cover him up and in a little while if he don't sweat give him another teaspoonful.' We did this and the child sweat and got well." Price .50

FAMILY *Poaceae*–Grass family
GENUS *Hierochloe* R. Br.–sweetgrass
SPECIES *Hierochloe hirta* (Schrank) Borbás–northern sweetgrass

Wild bergamot (Perfume)

HIDATSA NAME: *Maitapo'paduma* (This word has no
translatable meaning. GLW)
LOCAL ENGLISH NAME: horsemint (wild bergamot)
BOTANICAL NAME: *Monarda fistulosa* L.

Buffalobird-woman (vol. 20, 1916: 208)

I have told you how young men put braids of sweetgrass on the underside of their fans. I have seen horsemint used for the same purpose—bound on fans so that every wave of the fan brought some of the scent.

The men used to break off several of the plants at lengths of about six inches with the flowers attached and thrust them stems down under the thongs that bound the cloth about the handle of the fan. This they did to both an eagle-wing fan or to a tail-feather fan alike.

Young men also used to fan themselves with just a bunch of these plants. They were quite fragrant we thought. I hear that in South Dakota the plant grows much larger than it does on this reservation.

Horsemint thus used on fans was thrown away in a day or two when the plants had become dry.

In winter when we women went for wood, we would find these plants standing. We would crumple up some of the leaves and thrust them under our belts or into our mittens and thus made them smell very good.

We also mixed horsemint blossoms with sweetleaf [hyssop, *Agastache foeniculum*] blossoms to put in our pillows, which thus were made to smell sweet all winter.

I know of no uses of horsemint except for perfume.

FAMILY *Lamiaceae*–mint family
GENUS *Monarda* L.–beebalm
SPECIES *Monarda fistulosa* L.–mintleaf beebalm, Oswego-tea,
wild bergamot

Pine needles (perfume)

HIDATSA NAME: *maiditsitsi* (perfume pine) These were almost certainly ponderosa pine needles (*Pinus ponderosa*). There is a small stand of limber pines (*P. flexilis*) (Potter and Green 1964) in southwestern North Dakota, but limber pines have short needles, and Buffalobird-woman describes these needles as being long.

Buffalobird-woman (vol. 20, 1916: 350–51)

Use of pine needles for perfume

Young women used perfumes of stink grass [*mika iditsi*–meadow-rue (*Thalictrum dasycarpum*)] mixed with needles of perfume pine or *maiditsitsi*.

The pine needles were commonly used alone and not necessarily mixed with stink grass seeds.

As with the stink grass seeds, we often chewed up pine needles and spat or rubbed them over our clothing or in our hair. The usual way was to spit them into the palms and rub into our hair or garments—as on the sleeves of one's garment, for example. The pine needles were gathered, say, about two quarts, and put in a bag for winter.

Pine needles smelled good green or dry just as sweetgrass does, and they hold their scent several years. Young women and girls often filled a little bag of the needles and wore it suspended on a necklace on the back of the neck. In old days also we took very thin deer skin and made a long sack, sewed the whole length, to receive the needles. These tubes were filled with pine needles, and the sack was hung around a babe's neck like a necklace.

Young men also wore two skin bags filled with pine needles suspended on their hair switches. Each bag was about the thickness of a lead pencil and was bound together by a long string of

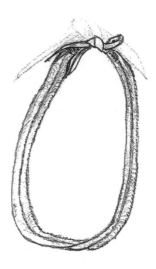

14. Pine needle necklace for a child. Drawing by Goodbird. (Courtesy of the American Museum of Natural History)

small beads wrapped spirally the length of the bag. Sinew thread stitched the bead strands together between the bags, separating them thus. The bags were about five inches long.

In the sketch by Goodbird of a perfume necklace for a child you can see that a seam runs down the middle making it in effect two bags.

The necklace hung rather loosely over the breast. A child dribbles at the mouth and for this reason the necklace was not drawn up where it would catch the mouth moistures. The seam down the middle was for neatness to make the necklace flat like a broad ribbon. Small thongs sewed to the ends bound them together.

Perfumes used in beds

Hidatsa beds took several forms depending on circumstances, but the beds in an earthlodge were hide-covered compartments tucked under the *atuti* (the lower, steeply sloping part of the roof) and offered the only "privacy" in the very open house. The "mattress" was a stack of bison hides with the hair left on. Coverings were the same and added as necessary. People slept in their clothes. Wolf Chief provides some vivid testimony to flea problems (as does Wilson).

Buffalobird-woman (vol. 9, 1910: 299)

In old times perfumes were kept in the bed. We used seeds that we gathered in this region, and we powdered up pine wood that comes floating down the Missouri River from Montana. We scraped the wood to a powder. Also we used pine needles which we got from the Crow Indians.

There were two or three kinds of native seeds we used which we pounded up in a piece of buckskin and chewed and spat over the bed.

Also we hung strands of sweetgrass over the poles (of the bed).

In old times we always paid a great deal of attention to perfumes and scents. Every man and woman had regard for perfumery. When a young man or woman went out, you could smell him or her quite a way off. Now-a-days our young folk don't care a bit for these things—don't care a bit to smell good.

Buffalobird-woman (vol. 16, 1914: 394–95)

When we made our pillows, before we sewed them up, we perfumed them. We took dry pine and shaved it down and rubbed the shavings all over the pillow cover. Or we took sweetgrass or

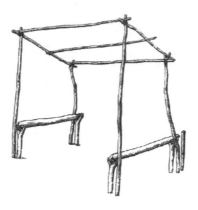

15. Bed frame (about 6 feet—1.8 m. square), which would have had a platform
of split-wood puncheons ("boards") and been covered with leather to form
a small compartment. Drawing by Frederick N. Wilson. (Courtesy of the
American Museum of Natural History)

other aromatic herbs and wet it and rubbed it over the pillow
cover, especially on the inside. When dry, the antelope hair was
put in and the pillow was sewed up.

When then we went into our beds they smelled sweet.

Beaver musk (Perfume)

Although obviously not a plant or plant product, somehow the topic of beaver musk came up, and Wilson decided to include this with his compilation of plant uses. Buffalobird-woman never says anything about it, but beavers (*Castor canadensis*) can be major predators of a corn crop, as serious as raccoons, deer, and ground squirrels. Where corn is accessible to beavers, river banks are littered with stalks cut and discarded by them. A beaver will quickly cut down many stalks, drag them to the river, and eat the maturing corn, letting the corn stalks float downstream.

Buffalobird-woman (vol. 20, 1916: 352)

Beaver perfume was also good. Just by the testicles of a male beaver and in corresponding position in the female are two glands or little sacks that smell good. A young man would take these out fresh, rake away the coals of a fire—so as not to burn the glands—and roll the glands in the hot ashes until half cooked. The mouth of the gland was then tied shut with sinew, and it was hung on a tree or pole to dry, which the outside skin did in about ten days. The gland itself was filled with oil and did not dry out.

The dried gland of a young beaver is small, and this a young Hidatsa man hung on his necklace as we call it—the beaded or quilled flap hanging down in front of his shirt from the neck and over the breast. Or he untied the mouth of the gland and squeezed a drop or two of the oil on his shirt. It made the shirt smell all summer.

Old men commonly dropped the gland inside of their paint bags. This scented the paint so that it smelled when the face was painted.

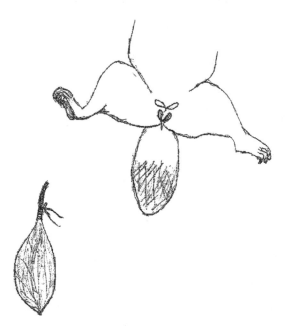

16. Location of beaver scent glands. Drawing by Goodbird. (Courtesy of the American Museum of Natural History)

The women of my tribe did not use this perfume, only the men. For women to use it would have provoked laughter. Among the Crows, women do, but they are fond of this perfume—inordinately so I think. To use it was not our Hidatsa women's custom.

SIX

Plants That Have Medicinal Uses

Big medicine

HIDATSA NAME: *xupadi itia* (*xupadi*, medicine; *itia*, big)
LOCAL ENGLISH NAME: (none given) [probably cow parsnip]
BOTANICAL NAME: *Heracleum maximum* Bartram

Buffalobird-woman (vol. 20, 1916: 333)

This plant grows always in the timber in the hills and has white blossoms. It is not very abundant, though hardly a rare plant either.

The Crows eat this plant. It is plucked when about 15 inches high. The fire is pushed away and the plant is laid on the hot ashes and rolled about with a stick so it will not burn. When [the plant is] well baked, the one cooking it takes the plant off [the coals] and peels off the bark or rind of the plant and eats it. It is the plant itself that is eaten, not the root.

The Crows are quite fond of this plant. I have eaten it once when visiting the Crow reservation, but I did not like it. It tastes rather sweet but does not lose the strong smell that marks the plant. We were on the Yellowstone at the time, and some Crows came into our camp to visit.

I do not know its use as a medicine. Only the Hidatsa doctors who use it can explain that to you.

It is the root that is used for medicine. I do not know if the plant itself is also so used, but I think not.

FAMILY *Apiaceae*–carrot family
GENUS *Heracleum* L.–cow parsnip
SPECIES *Heracleum maximum* W. Bartram–common cow parsnip

White and red baneberry (medicine)

HIDATSA NAME: *hudapi cipica* (medicine, black)
LOCAL ENGLISH NAME: black medicine (red baneberry)
BOTANICAL NAME: *Actaea rubra* (Ait.) Willd. [Red and white
forms are the same species.]

Buffalobird-woman (vol. 20, 1916: 314)

There are two kinds of black medicine. One had white and the
other red berries. But although I have gathered black medicine
all my life I have never myself seen the red berry kind until I saw
that specimen you have bought and now have in the chapel. I
knew about the red berry kind, but I never saw it.
[Wilson had his office/lab in Goodbird's church and sometimes
lived there as well.]
The root of this plant is sold for a dollar each here among the
Indians of this reservation.

The white berry kind grows down by the Little Missouri and
sometimes in the timber in the hills.

It was the medicine used in the River Ceremony and much
used in doctoring.

Both the white and the red kinds are quite scarce—the red
even scarcer than the white.

It was the root that was used.

Both varieties are equally valuable and used in the same way.

Gilbert Wilson (vol. 8, 1909: 134)

Black medicine

[A description from an inventory of items purchased for the
American Museum of Natural History.]

Of Maksipka [the person from whom he was purchasing the

item]. Black Medicine. Is specific against snake bite. If one carries a piece about with him he will not be bitten by snakes. If you chew a piece and spit it out in a snake's face, that snake will die right there. If chewed and spat and rubbed on one's clothes, snakes will smell it and want to get away. Is gotten on the reservation. Goodbird's mother [Buffalobird-woman] knows the plant. Mandan name is *niskadipsi* or medicine black. Hidatsa name is *hopadisipisa* [*hudapi cipica*]. Price .50.

[Note that Wilson's spellings of Hidatsa words evolved over time as he both studied the language and developed enough fluency to be conversant.]

FAMILY *Ranunculaceae*–buttercup family
GENUS *Actaea* L.–baneberry
SPECIES *Actaea rubra* (Ait.) Willd.–red baneberry

[According to O. A. Stevens (1963:142), the red-berried and white-berried are the same species, and although he says that the red-berried form is more common than the white, Buffalobird-woman had found that in her part of the state this was not so.]

Gumweed (medicine)

HIDATSA NAME: *mai'pucicipi* (no translatable meaning
so far as known. GLW)
LOCAL ENGLISH NAME: clap (gonorrhea) medicine
(gumweed or curlycup gumweed)
Grindelia squarrosa (pursh) Dunal

Gumweed is commonly found blooming along roads where almost pure stands of it can be found blooming in late summer. In the badlands of Theodore Roosevelt National Park, I have seen it growing along the edges of bison trails that are often ruts about 20 inches (about 50 cm) wide. It is a plant that favors recently disturbed soils, and competition from taller plants quickly eliminates it. The plant produces a sticky "gum" from which it gets its name.

Wolf Chief (vol. 20, 1916: 240–41)

This was one of my medicines—that is, it was a medicine that I had the right to use because I had bought the sacred right to it. I bought it from a Ree Indian [Arikara] named Mitotsadax as nearly as I can pronounce it, for the word is a Ree name. I do not know the meaning of the name. I bought it in the summer, and the very next winter that Ree man got sick and died. But before he died he sent for me and said, "I have taken sick because I sold you my sacred medicine, and I know I cannot therefore get well." I paid the Ree about two hundred articles—blankets, calicoes, otter skin, gun, etc., and made him a feast.

This plant has yellow blossoms, and the seed pods are very sticky with a gummy kind of juice that is in them. The plant is in flower now—in August.

[Wolf Chief's botany was imprecise. He is here talking about the calyx.]

To use this medicine, if one was bleeding from a cut or wound, I chewed up some of the seed balls and put the chewed cud on the wound, and the blood stopped flowing.

Also I would take two of the seed balls after they had gotten ripe and pound them up and put them in the bag in which I carried shells for my gun.

Now if I was out hunting and shot a deer or buffalo, if I hit it in the heart or liver, the animal died at once. But if I shot it through some nonvital part it would run away and die perhaps the next night: but if I had these powdered seed balls of this plant in my shell sack and then shot an animal in a nonvital place the blood was checked from flowing so that the wound swelled up at once and the animal died.

In laying the cud on a wound to stop bleeding, it did not matter whether the seed ball was green, ripe, or in blossom when chewed up. All were equally good. But only the ripe seed balls were put in with gun shells.

Once a boy was kicked on the forehead by a horse, and the flesh was cut, and the wound bled. I chewed a cud of seed pods of this plant and laid it on the wound and bound it up, and now the wound is healed and there is no scar that you can see. But I do not know of anyone else that used this plant.

Added by Goodbird: "My father told me also that a tea made of this plant's leaves and pods was good for gonorrhea."

(It should be noted that the local English name given in the beginning of the account of each plant in this list is usually, not always, the English name given by Goodbird. In the present case, the name is evidently suggested by the information given him by his father, Son Of A Star. GLW)

FAMILY *Asteraceae*–aster family
GENUS *Grindelia* Willd.–gumweed
SPECIES *Grindelia squarrosa* (Pursh) Dunal–curlycup gumweed

Purple coneflower (medicine, stimulant)

HIDATSA NAME: *xopadi taidia* (medicine, cold)
LOCAL ENGLISH NAME: cold medicine (black samson, but
it is usually called purple coneflower)
BOTANICAL NAME: *Echinacea angustifolia* DC

During the 1990s there was what might be called an "echinacea craze" in which numerous health benefits were claimed for the plant. Among these benefits was its alleged ability to "cure the common cold." This notion perhaps came from one of its names, "cold medicine." As can be seen below, the name was derived not from its capacity to cure colds but from the feeling of cold that it could induce. Coneflower plants were dug by the thousands and virtually disappeared from some prairie plots (and even private gardens and public parks). The more showy species (*E. purpurea* L.—the eastern purple cone flower) does not grow as a native in the Dakotas. One of Buffalobird-woman's sisters was named Cold Medicine, which could also be interpreted as Purple Coneflower.

Wolf Chief (vol. 20, 1916: 203–4)

Cold medicine plants grow on the prairies but not in the woods. The plant is found all over our North Dakota country.

I am not very well versed in all the uses of this plant because it belongs to the outfit of the River Ceremony, and I do not myself know the songs and rites of that ceremony.

But this plant has a particular use. It has a long root that goes straight down into the ground. When out on a war party, a warrior would dig up a section of the root of one of these plants—a piece about two inches long. This piece he would put in his bag or inside one of the extra moccasins he carried with him. Or he would tie it to the flap in front at the neck of his shirt that we call the necklace or on his back—anywhere as suited his convenience.

Well, sometimes we had to travel all night, and one would get tired and sleepy. When the warrior felt tired thus, he would chew a little piece of the root the size of one's fingernail and swallow the juice. The juice was very strong, and one could not take much of it. Also the warrior would wet the end of his finger in the juice of the chewed root in his mouth and with it moisten the upper eyelids and the inside of his ears and the paps of his breast. The juice felt cold and kept one awake and strong for the march. It has a rather stingy feel, like pepper but not so strong. When chewed and the breath was drawn into the mouth, the mouth felt cold.

This, I think, is why it is called by us cold medicine

FAMILY *Asteraceae*–aster family
GENUS *Echinacea* Moench–purple coneflower
SPECIES *Echinacea angustifolia* DC.–black samson, purple coneflower

"Medicine in the woods"

HIDATSA NAME: *midamahu xupadi aku iditsi saki* (woods in, medicine, kind of, smells, good = medicine in the woods that smells sweet)

LOCAL ENGLISH NAME: medicine-in-the-woods (unidentified). I am going to hazard a guess here and identify this plant as sweet cicily, which is known as longstyle sweetroot to the USDA botanists. This fairly common woodland plant seems to me to be what Buffalobird-woman is describing. The seeds smell like anise, and in folk medicine the root is used.

BOTANICAL NAME: *Osmorhiza longistylis* (Torr.) DC.

Buffalobird-woman (vol. 20, 1916: 334)

This is a medicine plant. Our Hidatsa doctors should teach our young people how to use this plant, but our Indian doctors are dying off.

I do not know many of the medicines. Our native doctors used different kinds of medicines for different kinds of sickness. I do not know what this plant was used for.

Our young people are growing up without knowing the lore of our native doctors, and they do not learn of their medicines.

I never passed into all the ceremonies and do not know the rites of any except those into which I have passed. I can tell nothing more about this medicine.

This medicine-in-the-wood plant grows in the timber in the hills.

We used to gather our medicine plants before they got as mature as the specimen you have here.

My mother used to gather up black medicine [red and white baneberry] and medicine-in-the-woods for her own medicines. The root was used.

FAMILY *Apiaceae*–carrot family
GENUS *Osmorhiza* RAF.–sweet root
SPECIES *Osmorhiza longistylis* (Torr.)–longstyle sweet root
and also known as sweet cicily

Poison ivy (medicine–wart remover)

HIDATSA NAME: *midapoti* (*mida*, wood; *apa*, leaf; and *oti*, ripe. So
called because the leaves in autumn turn red like the color of ripe fruit,
while the leaves of most other plants in the woods turn yellow. Goodbird
seems to say that both *apa* and *midapa* are used for leaf. GLW)

LOCAL ENGLISH NAME: poison ivy

BOTANICAL NAME: *Toxicodendron rydbergii* (Small ex Rydb.) Greene

Just how people like Buffalobird-woman's husband who used the
plant avoided severe poison ivy rashes I do not know. All parts
of the plant are toxic. Poison ivy is an extremely adaptable plant,
and I have seen it as vines 3 inches thick climbing out of sight
in trees, as 4 to 5 foot tall free-standing plants, and as a plant
no more than 2.5 inches (about 6 cm) tall growing at the edge
of a butte in North Dakota. It does indeed stand out in the fall
when many poison ivy plants tend to be a fairly bright red. As
Buffalobird-woman notes, there are virtually no woody plants
that turn red in the fall in that part North Dakota.

Buffalobird-woman (vol. 20, 1916: 338–39)

We used this plant for only one thing—to take off warts. For
example, I once had a wart on my nose. My husband took a piece
of the woody stem of the plant and broke it in two. It was in the
spring when the sap was flowing. He pinched off the leaves at the
top and, in the wound thus left, a little milky juice exuded which
he increased by drawing the section of stem between his thumb and
finger. The little drop of juice thus collected on the end of the stem
he touched to the wart. It was sore the next day but not irritating
to me. In two or three days my husband applied the same again.
Then, after an interval, he made a third application. The wart got
sore, but the wound was not at all irritating to me. The juice rotted the wart, roots and all. After five or six days, the sore healed.

The wart did not grow again all summer. But the next winter it returned. I tried the remedy again the next spring, applying it three times. This took the wart out, roots and all.

This use of the plant for removing warts was very general in the tribe. Three applications, I think, were usually made. If a man used the plant and the wart returned, he tried it again the next year as I did.

Boys used to apply the plant to the wrist in the same way as described, making little dots on the skin. They did this in order to be tattooed. The juice made a mark wherever it touched the skin, but this only remained for a few days. No permanent scar remained.

When we went into the woods we used to show the plant to our children and say, "Don't touch this!" For we feared some child might break off the plant and chew it.

I do not remember that we thought merely walking through the growing leaves of this plant or touching them would poison us. We did not regard them as especially dangerous. However, I remember that once some of these plants were in my garden. I pulled them up by the roots as I dug them out, and afterwards sores came out on the back of my hand.

FAMILY *Anacardiaceae*–sumac family
GENUS *Toxicodendron* P. Mill.–poison oak
SPECIES *Toxicodendron rydbergii* (Small ex Rydb.) Greene–western poison ivy

Unknown grass (medicine—treatment of eye ailments)

HIDATSA NAME: *maictixoxi* (*maicta, eye*; and *ixoxi*, scratch)

LOCAL ENGLISH NAME: (none given)

[This may be American sloughgrass (*Beckmannia syzigachne*
(Steud.) Fern.)]

Buffalobird-woman (vol. 20, 1916: 337)

This is a grass, the leaf of which has a rough upper surface. It
grows in level places and is very plentiful on the meadows.

It is used to scratch the eyelid on the inside—to draw a little blood. But I do not know whether the side of the grass blade
or the edge is used. Many Growths is an eye doctor and would
know. Such knowledge was bought and sold by the eye doctors.

The remedy was in use from very old times as I have heard,
and it was used for old and young. Eye sickness is very common
among us.

(Goodbird adds that he has heard that the eyelid is turned
wrong side out—the upper eyelid only. It is scraped with the
rough surface of the grass blade held doubled between thumb
and forefinger in a loop. Whether edge or broad surface of the
grass is used he is not quite sure but thinks latter. The eyelid is
scraped until blood comes.

Mr. Huber gives corroboration of the above and says the broad
rough surface of the doubled grass blade was used. Treatment
was for trachoma and granulated lids. As a rule only the upper
eyelid was scraped for fear of hurting the eyeball. GLW)

FAMILY *Poaceae*–grass family

GENUS *Beckmannia* Host–sloughgrass

SPECIES *Beckmannia syzigachne* (Steud.) Fern.–American
sloughgrass

Peppermint (medicine)

HIDATSA NAME: *hicu*
LOCAL ENGLISH NAME: peppermint (wild mint)
BOTANICAL NAME: *Mentha arvensis* L.

Buffalobird-woman (vol. 20, 1916: 193)

This plant is a medicine plant used in doctoring women thus: If a woman gives birth to a child, she drinks water with this peppermint in it. It is not boiled; the plant is dipped in cold water and drunk. It stops the flow of blood.

The woman does not drink before but after the birth because its purpose was to stop the hemorrhage. It is very commonly used on this reservation now after childbirth.

This use of the plant was not safe for any other hemorrhage—as from a wound or from the nose.

I know of no other use ever made of the plant. It was not burned for incense.

(Specimens collected were found in Long Bear's wife's potato patch. To have some handy she said she had transplanted some of the plants, and they had spread. She knew of no transplanting in old times for like purpose. GLW)

I never knew of these plants being transplanted in old times as you say Long Bear's wife has done. I remember someone planted seed of the prairie turnip, but they did not grow and nothing came of it. I never heard of any attempts to grow any other wild plants. Besides, there were plenty of all wild plants near the village anyway, and we felt no need to plant and raise them.

FAMILY *Lamiaceae*–mint family
GENUS *Mentha* L.–mint
SPECIES *Mentha arvensis* L.–wild mint, field mint

SEVEN

Plants Used for Fiber

Dogbane (utilitarian—fiber)

HIDATSA NAME: *mida wahu acu* (*mida* woods, *wahu* in, *acu* snare)
LOCAL ENGLISH NAME: snare-in-the-woods-plant
(spreading dogbane but often called Indian hemp)
BOTANICAL NAME: *Apocynum cannabinum* L or
A. androsaemifolium L. (probably *A. cannabinum*).

Spreading dogbane or Indian hemp has milky sap like milkweed, but the flowers are somewhat different, and the pods are long and pencil thin. In the fall the foliage turns a bright yellow, making the plants stand out from the surrounding vegetation.

Wolf Chief (vol. 20, 1916: 273–90)

This snare-plant grows in wet places and in low-lying grounds where water lingers long after the rains. In such places along the edge of the Missouri timber and in the hills, it grows thickly. It grows only in the woods, never out on the prairie. It grows all sizes—from 2 or 3 to 4.5 feet (60–135 cm). The flower is of the same color as the plant, and when it opens has a kind of cotton inside. [Wolf Chief mistook the pods for the flowers, which are small and white.]

The word *acu* must be a very old word, for it is spoken of in very old ceremonies, as in the Black Bears' Ceremony, which we recognize as the first or beginning ceremony.

In this Bears' Ceremony there is a place in the ritual where the "Big Medicine" says, "I am a large (great) power snare." Also in this ceremony they peeled off the bark of this plant and made snares. I have seen my father, in order to prepare for the eagle hunt, go out in winter and gather this snare-plant for snares. He would break off the plant and break and work and twist it in his hands, pulling off and throwing away the coarse shreds and bark and retaining only the fine hair-like shreds for the cord he wished to make.

The plants were perfectly dry and dead by the end of September, and then was a good time to gather them. They were gathered by men, not by women. I have gathered the plants myself.

I thus gathered the plants. I broke off the blossoms [the pods] and threw them away. It was believed by us that if any of the cotton in the blossoms got into one's eyes he would go blind. Then I grasped the top of the plant and kicked it above the roots with my foot to break it off. However, some shreds still hung fast, and I twisted the plant to free it wholly from the roots. Down near the ground the plant was usually a bit green even after the top had died.

I now bit the stalk flat with my teeth, beginning with the root and shoving the stalk along in my mouth as I bit. I next opened the stalk the whole length along one side. Then I would break back a piece about two inches in length, peel out the pith part, and throw this away. Then another two inches, and so the whole length of the stalk.

We sometimes did this as we gathered the plants; sometimes we brought a quantity of the plants to the lodge and prepared them there.

I next shredded the plant—working it with my thumb, bending it and rubbing the rind or bark left after removing the pith, thus shredding the fiber like sinew. There would be a little rough rind left on the outside, but it amounted to nothing and easily fell off during the shredding process. All this, as I have said, might be done in the field where the plants were gathered, or it might be done at home in the earthlodge.

To make a cord, I gathered in my hand as many fibers as would make a cord the thickness I wished, and then I tied them in a knot at the root end. The fibers I now divided into two equal parts. I bared my right thigh, for this purpose pushing my legging down over my knee. I spat on my right palm, pressing the two strands of fibers smartly against my thigh. I twisted them separately with a forward stroke, and then with a smart half stroke backward

I twisted them together into a cord. This would make a section an inch or two long. The process I kept repeating.

When either of the two strands began to show signs of thinning out, I added more of the shredded fiber, which I twisted slightly with my thumb and finger upon the thinned strand—just enough to make it grip and hold. New fiber thus added to the thinning strand was always added from the root end.

Some men were good cord makers—or as we said, good snare makers. One such was named Has Many Antelopes. Men paid him with a knife or some other article to twist cord for them. They would bring the fiber to him all ready to twist. But I never knew of snare fiber to be laid up for a year's use as women laid up bark strips for baskets.

Whether the fiber of plants gathered up in winter months was good I do not know. I always gathered it up for my own use in the fall. Nor do I know of anyone storing away cord for winter use or for later labors.

In strength I think snare-plant was equal to sinew. It was, I believe, of just the same strength.

The heaviest cord that I ever saw made of snare-plant fiber was for a fish line, and some of our fish lines were quite coarse. This line had a rather large hook on it—a steel hook gotten from the traders. For a sinker one used a small stone, elongated, and of a shape so that it would not fall off the line when tied about the middle. This fish line was about as heavy as the sinew cord on the rabbit snare I have made for you.

It was our custom to consider the snare-plant sacred. We might have used the fiber cord for a great many other purposes, but we were rather afraid, for its recognized use was to make snares to catch game—we counted a fish line as a kind of snare.

The root we considered a medicine, and it was used in the Bears' Ceremony. Also if anyone was hurt in the eagle hunt by the eagle's claws, some of the root of the plant was applied, and the wound soon healed. If the glands of a man's neck swelled up, the roots of some snare-plants were pounded fine, boiled, and the water was put on the glands.

Charging Enemy once told me that a little bag of the roots pounded up, if smelled, would make a man sneeze, and that thus used they would make weak men strong.

Also if any piece of a snare-plant was thrown outside the lodge, and someone went outside and carelessly urinated on it, he would not thereafter be able to urinate, for the power to do so would be taken from him. Such was our belief, and it was for this reason that we commonly shredded and worked the plants out in the field where we gathered them rather than in the lodge.

As I have said, we regarded the plant as sacred or magic. It was to be used to make cord to catch birds, beasts, or fish. When it was used in sacred ceremonials, these too had to have something to do with the snaring of game.

Snare-plant cord was used for fish lines, for nooses to snare prairie chickens, weasels, rabbits in the woods, gophers, and to make ceremonial snares for use in the eagle hunt.

Snare-plant cord was also used for binding cord if used on something that had to do with snaring game. Thus when a rabbit snare was made, if the wooden handle was split to put in the noose, the wood was bound together again with the cord in three places.

Deer snares

Also I have heard that in old times our people twisted snare-plant fiber to make nooses or snares to catch deer. But deer are strong animals, and they used to break the snare-plant ropes that made the nooses and escape. But certain men in the tribe had sacred power or sacred right to make imitations of these plant fiber snares out of green buffalo hide. These were stronger, and with them they caught deer. However, deer were strong, and they were also hard to catch, so not many were taken. The snares were set at night, and yet when the deer came, they would stop, turn, and go back, being keen to suspect something.

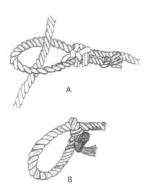

A

B

17. A model of a deer snare made by Wolf Chief. *A* shows the snare in the open position; *B* shows the snare knot tightened. Drawing by Goodbird. (Courtesy of the American Museum of Natural History)

For this reason, our people used to choose a timber or wood that ended in a point, and here on the point, where the forest narrowed so that there could be but one path, they set the snare. They then went into the wood and frightened the deer, and the scared animals would run out into the point and right into the snare that was set there. If caught by the neck, the deer died very quickly. The animal would cry, "Hwa-a-a" or a sound something like that, and die.

My father told me of this. He never snared deer himself but saw it done.

The snare used for catching deer was made with a kind of double loop, or noose, as shown in this drawing. The object of this was that when the deer pulled the greater noose taut about his neck, the second noose held the first noose tight.

In my youth, the tribe had gotten guns and no longer caught deer with snares.

Rabbit snares

We made rabbit snares of snare-plant fiber, or of sinew cord if we could not get snare-plant. However, if we could get plant fiber, we preferred to use it.

I have caught rabbits in sinew snares. They were like those I made for you for the museum and the university.

I used to see my father making snares to catch rabbits, and so I knew how to make them myself. I do not now remember very clearly the details of his trapping rabbits with snares, except that he often made snares in the woods and caught rabbits with them. He often did this.

I began snaring rabbits myself when I was about fifteen, I think. We were in camp in the hills timber. Hunting had not been very good, and we were hungry in the camp. I remembered how my father made snares to catch rabbits, and I thought I would make some snares and catch some of the many rabbits that I saw were about. I forget now, but I think that I used four or five sinews, and I used boxelder for snare handles because boxelder sticks have large pith cores and so are easily hollowed out. Also I cut some large-sized rose bush stems and used them for snare handles.

In the evening I took out the snares and looked for places to set them. The rabbits made many trails through the bushes. Always I put two sticks in the snow with the snare between them. The other end of the sinew noose I tied to a bush or small tree.

I set two or three snares by the side of a ravine and came home.

In the morning I went out and found I had caught three rabbits. They were dead.

I skinned the rabbits when I got back to our tent. The first I stripped off, cutting a little slit between the fore legs; and putting in my fingers I drew the skin to right and left. I threw away the skin. Then I got to thinking. It was cold, and the rabbit fur would be nice to put in my moccasins. The rest of the rabbits I skinned by slitting along the length of the belly and taking the skin off in one piece.

Some of the rabbits were fat, but one or two were not. These rabbits ate bark and so were good to eat, tasting like antelope.

I gutted the rabbits, and my sister Buffalobird-woman boiled them. I helped her boil them also. She and her husband and I ate

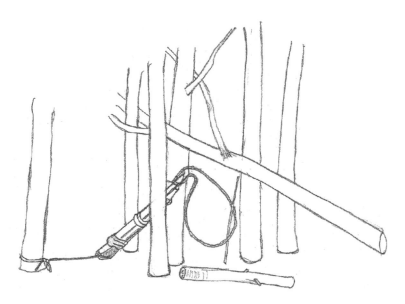

18. A rabbit snare. The upright stick in the loop keeps the loop open. Drawing by Goodbird. (Courtesy of the American Museum of Natural History)

the rabbits; also I think someone else came in and ate with us, but I forget who it was. At this time all in camp were hungry.

The drawing shows how a rabbit snare was set. It was set the same way whether over a rabbit hole or on a rabbit trail. A stick was laid under the noose as shown, to make the rabbit jump over it into the noose. The noose should have the breadth of two fingers above the ground—quite enough, because a rabbit's forelegs are short. The end of the cord trailing from the handle was bound to a bush or sapling. The twigs on which the snare was set were all small so that the snare was easily dragged loose by the rabbit when caught.

As I have said, I set rabbit snares in a rabbit trail or else before a rabbit hole because gray rabbits lived in holes in the side of rising ground, and jack rabbits also dig holes. If I put a snare in a rabbit trail I chose a place where the bushes were thick, or else I made the place so by thrusting little branches into the ground on either side. Besides the stick at the bottom, I thrust one across at the tops.

Use of the rabbit skins

I always used dry grass instead of stockings as we now wear after whiteman's custom. That is, in cold weather I put dry grass in my moccasins to keep my feet warm. Now I took those rabbit skins and worked them gently with my hands to make them soft. Then I put dry grass in my moccasin, and wrapped each foot and tail end of the skin over my toes, lapped up on either side. It was midwinter and putting my foot in moccasins with grass and rabbit skins kept them warm when I slept. I never knew anyone else to do this—I just thought of it myself.

Also I put those rabbit skins inside my shirt over my chest when I went after the horses because it was very cold. I did this just this trip, I mean, for I was a boy, and it was very cold weather.

Other incidents of the hunting trip

Now I will tell of this hunting trip as much as I can remember. We had gone into our winter village, camping some 10 miles or more up the Missouri from Fort Berthold. We found poor hunting that winter. There were no buffalos, and the people were hungry.

Some of us went up the river hunting—Heart's Enemy, Crow Flies High, Wolf Goes With The Wind, Bear's Necklace, and others; we were about a dozen tents in all. I went with my sister Buffalobird-woman and her husband—her first husband, Magpie.

I cannot remember very well—I was fifteen at the time—but we were camping in some timber in a ravine near the Slides—opposite Independence and about 7 miles from the Missouri. We got water at a spring. There were many beavers in the brook that flowed from the spring. We had brought with us boiled green corn, dried squash, and other dried vegetable foods from our gardens, but no dried meat. Buffalobird-woman and her husband and I used one tent.

The men sent out two young men who went out some 10 miles away and looked around from the tops of the hills for buffalos but did not spy any. The next day they went in another direction but found none.

They were sent out again, or two young men were—Bear's Necklace was one I remember. He saw one buffalo and killed it. Then we all ate our fill, for the carcass was cut up and distributed through the camp.

When Bear's Necklace killed the buffalo, he piled the meat up in a pile, and the next day the other men went out on horses and fetched the meat in.

It was very cold. Doubtless Bear's Necklace wore mittens. He killed the buffalo with a gun.

Protection from wolves and frost

Now perhaps you think a dead buffalo so left out in the weather would be frozen the next day. But we Indians knew how to fix it so that it would not freeze.

When the carcass had been cut up, we looked for a little hollow, or a place where the buffalo had been rolling. From such a hollow the hunters cleaned out the snow with their feet till they could see the grass. Then they took the chewed up grass in the buffalo's paunch—which chewed up grass is nearly dry—and spread it on the ground in a kind of pile. On this we piled the cuts of meat and covered all with the skin, fur side turned to the weather.

And now the hunters covered skin and all with snow, scraping it up with their feet, until about 2 feet of snow covered the meat pile.

The next morning the meat would be found unfrozen—or even at the end of two days. And this in the coldest weather.

Also when we hunted, we Indians used to tie cloths to our gun rods which we thrust in the snow beside such a pile of meat; this was to frighten the wolves away. And we would urinate on the ground nearby. Sometimes we chewed black medicine [red baneberry, *Actaea rubra*] and scattered the chewed bits around. Wolves then never came near the meat.

Later on, I know, we would take a pipe and clean it out, scattering the ashes around the meat pile, or we would chew some tobacco and spit it on the ground about the pile. But this was a later custom.

Camp is removed

Crow Flies High was leader of our camp. We wintered at the place where I caught the first rabbits—or at least we stayed there quite a while. When our corn was all eaten, we sent a party of about ten men back to the village for more. They came back bringing with them my friend Garter Snake. Also we heard now that some other of the villagers had gone up the river to White Earth Creek, that they had found many buffalos, and had sent back dried meat to the village.

"We must go over there to them," we said, "for there are buffalos there." All the camp agreed to this and that they would move the next morning. "We will camp halfway there tomorrow evening," we said. "Then the day following we will reach their camp!"

We started the next morning. The snow was very deep and the people led their horses and walked. Before sunset we camped in some woods near a creek. I went into the timber and found many rabbit tracks there and some rabbit holes. I set my snares.

That night we went to bed, and I awoke early the next morning. Then I heard a gun fired, and someone said, "There were two buffalos out there—a hunter killed one; the other got away." Then I saw a man running and a buffalo also running slowly, for it was wounded. The buffalo ran along one side of a ridge of ground. The man following ran along the other side of the ridge, and, when the buffalo came near him, shot it. The two buffalos had in some way gotten lost from the main herd. We now had plenty of meat.

I went out to the slain buffalo, and when they cut up the carcass and took out the liver and kidneys and paunch I ate a small piece of each. Our Indian people did not waste anything; even the blood was saved in pails.

When I got back I went out to see the rabbit snares I had set the night before. I found I had caught two rabbits. They were gray ones.

We used to have three kinds of rabbits about this reservation. Jackrabbits, small rabbits that turned white in winter, and gray rabbits that stayed gray all the year round. About twenty years ago the gray rabbits began to disappear. Whether they died out, or what, I do not know.

The two rabbits I had caught I skinned, and we ate them, but the skins I threw away.

Our camp had sent two men ahead to go to the White Earth camp. These two men now returned saying, "We did not go to the White Earth camp for when we came to some hills we spied many buffalos ahead." So they returned with the good news.

We moved camp this same morning; we had eaten up almost all the two buffalos killed—they were only yearlings. We went toward White Earth Creek, finding tracks of horses. Then we came to a hole in the ice of the creek where horses had been watered through the ice. We followed the tracks and came to a place in the timber where there was an abundance of meat—for the camp there had made a good kill. Meat, fresh and dried, was piled upon stages; dried and frozen meat both.

We found our friends and camped. Our friends gave us fresh and dried meat to eat. They called us into their tents and gave us roasted buffalo to eat. When we had eaten in one tent, they would call us into another to eat again. We had been on short rations, and now we all ate ravenously, so that that night we all got sick, vomiting and physicking at the bowels because we had eaten too much.

Describes camp

This camp at White Earth Creek was all of tents except two lodges that were built just like eagle hunters' lodges with logs stood up on end all around them. This was for fear that enemies might come and shoot into the camp. If enemies had come, all the people would have run into those two lodges—which are warmer than tents. The two lodges were each about twenty feet across the floor, and they stood a few yards apart. The tents were grouped irregularly near the two lodges. I think there were five tents besides the two lodges before our own party joined them. [Wolf Chief is clearly talking about tipis rather than tents.]

Life in camp

The tent my sister and her husband and I camped in was of good size. I think we tied six poles on each side of a horse as I remember. We had eight horses along with us; eight horses that belonged to our family, I mean. Other families had eight to ten horses with them also. This was because we had hopes that we might kill lots of buffalos and fetch the meat home. Some had travois; at least I remember a few did, though our family had none with us. But we did have a travois basket, which we tied on the tent poles, thus making a kind of travois with it.

Beds in our tent were made by first putting down dry grass on the ground. Over this, we put robes and saddle blankets. Then we put our own robes on top of these. A log fenced the bed from the fire. We were careful to clear away any hay or straw (dry grass) that might fall between the bed and the fire.

We stayed about a month at White Earth camp hunting and drying meat and hides.

Then my sister asked me to go back to the village with meat. Garter Snake had camped with me in my sister's tent. My sister also gave him a good share of meat to take home, and he and I journeyed home together, taking our meat with us.

Life in the hunting camp

My sister was a very industrious woman; there were very few women in the tribe that at all compared with her as a worker. She was busy all the time, with the fire outside, drying meat and dressing hides. She dried the meat by the fire. She put up a drying stage and built a fire under it so that the meat, or some of it, was cooked; and all of it was dried in the smoke.

At evening, after the meal, the men would gather together in one of the two hunting lodges I have described and tell stories of old times or of their own time. The others would listen and smoke while one talked. Women did not enter when the men were gathered there.

Sometimes in the day, women might ask permission to go in this lodge and scrape hides; sometimes on very cold nights some people slept there.

But no family or families lived in either of the two hunting lodges regularly because they belonged to the whole camp. Anyone who cared to might go in there and sleep, but it was not home for anyone.

We had no special name for either of these two hunting lodges.

Snaring ducks

And now returning to the subject of snaring—my father used to snare ducks; but I never saw him do it, nor did I ever snare ducks myself.

However, I know how it was done. He used to snare ducks on the lake (slough) at Shell Creek—the place where the Missouri overflows and makes a lake when there is high water. He used to snare them in the fall—about September when the young ones are flying. He did not catch them in great numbers.

My father got sticks a little smaller maybe than my little finger and about 4 feet long. The snare was a noose bound about two-thirds of the way up the stick. The longer end of the stick was to be thrust down into the mud.

The ducks made little paths between the reeds. These paths were water paths, paths in which the ducks swam. My father would thrust a stick down into the mud so that the noose just touched the water. The noose was like the one I described as used to snare deer—with a double knot—a choke knot. The noose hung about 3.5 inches in diameter and just in the very middle of the run or path of the fowl.

I remember he used to go out with about ten snares and sticks at a time. He would wade into the water and set them. Sometimes he caught a duck in nearly every snare. He caught them of all sizes, young and old ducks alike.

Small bird snares

Snares for small birds were just ordinary running nooses, not the kind that tightened at the knot like those for ducks and deer.

FAMILY *Apocynaceae*–dogbane family
GENUS *Apocynum* L.–dogbane
SPECIES *Apocynum androsaemifolium* L.–spreading dogbane

FAMILY *Apocynaceae*–dogbane family
GENUS *Apocynum* L.–dogbane
SPECIES *Apocynum cannabinum* L.–Indian hemp

Upright sedge (utilitarian—fiber)

HIDATSA NAME: *Midapa ita wika* (beaver's grass)
LOCAL ENGLISH NAME: beaver grass (upright sedge)
BOTANICAL NAME: *Carex stricta* Lam.

In Buffalobird-woman's description of the habitat in which one can find beaver grass, she says that it was found "where beavers have their dens." Along rivers and creeks that have the potential to overwhelm and wash away beaver dams, beavers dig tunnels into the banks. These sites are often where river currents are slight. On smaller creeks, if frequent flooding is not a problem, beavers cut and trim trees to make dams to create more or less permanent ponds around which aquatic and semiaquatic vegetation will grow.

Buffalobird-woman (vol. 20, 1916: 270–71)

This grass was called beaver's grass or beaver grass because it was found growing where beavers have their dens; I do not know why—but this I have heard. It grows in the woods along the Missouri River in wet places.

It was from this grass that we made strings for stringing dried squash.

The grass was cut with a knife and brought to the lodge. My father used to do this for us. He brought it in bound up in a bundle and packed on his back. The bundle was big enough to fill his two arms.

Sometimes my mothers brought in the grass; sometimes my father, as I said, went alone. Sometimes he and my mothers went out together. Care was taken not to break the grass. It was laid in the bundle stem and stem, and top and top, very carefully.

19. Carrying a beaver grass bundle. Drawing by Goodbird. (Courtesy of the
American Museum of Natural History)

This is a sketch of a bundle such as I saw my father bring in. Note
that the stem end is toward the left. In packing a bundle on the
back, the carrier always adjusted it so that the trailing part was
toward the right.

The grass having been brought in, my father took it up on
the corn stage and spread the grass out evenly, three or four
inches thick, over the floor of the stage. He laid the grass stem
and stem very carefully and weighed it down with sticks so the
wind would not blow it about. Here it was let dry, which it did
in three days.

My father then fetched in the grass into the lodge and laid it
back of the beds in the *atuti*—that part of the lodge under the
descending roof where we stored things.

I would take a small bunch of the grass and soak it overnight
in a pail. In the morning my two mothers would make strings,
and sometimes I helped.

Beaver grass differs from other grasses in that a stalk grows
up and branches into four or five leaves. It is by this peculiarity
that we could tell the grass.

20. Needle for stringing dried squash, with a beaver grass cord. Drawing by Goodbird. (Courtesy of the Minnesota Historical Society)

Gilbert L. Wilson (vol. 29, 1916: frame 0286)

Making a grass cord for stringing squash

August 16, 1916. Been out all morning watching Owl Woman, Mandan, slicing squashes and twisting a cord of grass on which to string the dried squash slices.

The cord is of beaver grass as Owl Woman calls it. The grass has been plucked carefully, dried, and then dampened.

She begins by tying together at the root ends two strands of two or three stalks (each) and over the knot so made she ties the end of her moccasin string which she loosens for the purpose. Then with her right hand she twists the right-hand strand several times toward the right; then shifts or loops the twisted strand back to the left and over the other and untwisted strand. The latter now becomes the right-hand strand, and this she twists to the right and laps it to the left over the first strand which now again becomes the right-hand strand. The latter is seized in the right hand, twisted to the right, and looped over to the left as before. And so on. The effect is to keep twisting each single strand to the right while the two strands doubled are by the lapping automatically twisted to the left.

As the cord proceeds, the strands are each lengthened as required by adding other grass stalks laid always with root ends toward the moccasin.

As the cord length accumulates, Owl Woman wraps the completed cord about her foot to keep it out of her way and so that she may draw the untwisted ends of the strands taut as she works.

The untwisted ends also are thus kept at a convenient length for her labor.

FAMILY *Cyperaceae*–sedge family
GENUS *Carex* L.–sedge
SPECIES *Carex stricta* Lam.–upright sedge

Grasswork ornaments on leggings

UNSPECIFIED SPECIES

Isokikuas or Leader (husband is Lance Owner)
(vol. 10, 1911: 283–85)

Lance Owner's wife had contracted to make a pair of leggings to accompany the quill wrought shirt she sold the museum two years ago. She gave the following information concerning them:

"I began the quillwork on these leggings in midwinter of this year and finished them in June, working at odd times. The pattern is old-time. A diamond pattern is Sioux. We Hidatsas used only bars as here. But in this pattern the stripes, you note, are slightly aslant. We sometimes wrought them on so that when the leggings were on the wearer's person the stripes would be perfectly horizontal.

The color arrangement was not always the same. Sometimes the background which I have here worked yellow was worked in white quills. One's taste guided in these matters.

In this pair, the yellow band running up the side of the legging is broken by four stripe patterns in purple, white, and green. This was the customary arrangement. I have seen pairs with five or three stripe patterns but never six or two.

The stripe patterns are made each in five bars, because I think that the most tasteful. Years ago I have seen these in two bars, or three bars, and even in four bars.

The little checkered stripe on the bottom of the legging I put there because I thought it tasteful. It has no meaning.

In my time, we have used only steel needles to work porcupine quills. In old times I hear we used sharp elkhorn awls. We wetted the end of the sinew with which we sewed and let it dry. It thus made a point that would go right through the hole punc-

21. A bone awl, typically 4–5 inches (10–12 cm) long. Drawing by Goodbird. (Courtesy of the Minnesota Historical Society)

tured in the skin by the awl. Awls were also made, I have heard, from Rocky Mountain sheep horn.

I saw the awl my mother used when I was young, but I never saw her use it. My mother said she used it on porcupine quill work.

In old times we used three different kinds of material to make the bars of the stripe patterns: porcupine quills, bird quills, and grass roots.

Grass roots were used only to make the black bars because they are not very strong and are easily broken.

We did not color grass roots for they are black of themselves. We put moist earth over them, took them out wet and pliable, and sewed them on the skin. Then we pressed the sewed root flat with the fingernail. [See Buffalobird-woman's comments in the section on buffalograss.]

We still used grass roots as I have shown in the specimen I made for you, but we do not use them very much. You can dig grass roots almost anywhere on the prairie, but along the edge of the high banks of the Missouri we can find the longest and strongest roots.

When I do porcupine quill work, I also put a little moist earth over the quills to make them pliable. I take enough quills out of my case for the work immediately at hand and cover them with just a little moistened earth."

Goodbird interrupting: "But my wife just wets them with her mouth."

Leader continuing, "I use moistened earth because it was the custom of my ancestors."

EIGHT

Plants Used for Smoking

Tobacco 9a

HIDATSA NAME: *opi* (tobacco)
LOCAL ENGLISH NAME: tobacco
BOTANICAL NAME: *Nicotiana quadrivalvis* Pursh var. *bigelovii* (Torr.) DeWolf

As recently as the mid-1990s, this species was thought by some botanists to be extinct. At that time I had it growing in my garden, having obtained some seeds from an Iowa schoolteacher who grew Indian crops. I once received a phone call from a taxonomist who had somehow heard of my garden and tracked me down. I assured him that it existed and sent pictures.

It readily self-seeds and is one of the first plants to germinate in the spring, usually by the thousands. The seeds are tiny and germinate best if just "dusted" over well prepared soil and smoothed over with the hands. Planting in rows as Buffalobird-woman describes just makes it all the more difficult to get it to grow. It can reseed itself about twice before it has exhausted the soil and the plot has to be established elsewhere.

Although I have done no qualitative research regarding its effects on plants growing in its vicinity, it does seem that corn hills adjacent to tobacco do not do as well as those farther away. The tobacco has a strong and distinctive smell that transfers readily to anyone handling the plant.

Plants typically grow to about 30 inches (75 cm) tall and flower profusely. The flowers are white and tubular—about an inch (2.5 cm) across. The flowers attract sphinx moths (*Manduca sexta*), which apparently do much of the pollination. The plants are also attractive to the sphinx moths as a place to lay their eggs. The resulting caterpillars grow into 4-inch (10 cm) hornworms. The hornworms can denude a tobacco plant in a single day. So the plants can provide food for both the adults and the young of the sphinx moth, and the gardener must be ever vigilant. I think that these are the insects that the Hidatsas call "tobacco blowers."

This tobacco species is originally from California and adjacent states (USDA nd) and was probably traded into the Missouri villages up the Columbia River and down the Missouri by the same route that dentalium shells, valued for decorative purposes, made it into the region.

The two accounts that follow are the perspectives of Buffalobird-woman, who, because she was a woman, did not raise tobacco but was very familiar with its propagation and use, and that of her brother Wolf Chief, who maintained a tobacco garden and prepared tobacco for smoking for years. Buffalobird-woman's account was the basis for a chapter in Wilson's dissertation (1917: 121–27). Spelling and punctuation are as published.

Buffalobird-woman (Wilson 1917: 121–27)

Tobacco

Tobacco was cultivated in my tribe only by old men. Our young men did not smoke much; a few did, but most of them used little tobacco, or almost none. They were taught that smoking would injure their lungs and make them short winded so that they would be poor runners. But when a man got to be about sixty years of age we thought it right for him to smoke as much as he liked. His war days and hunting days were over. Old men smoked quite a good deal.

Young men who used tobacco could run; but in a short time they became short of breath, and water, thick like syrup, came up into the mouth. A young man who smoked a great deal, if chased by enemies, could not run to escape from them, and so got killed. For this reason all the young men of my tribe were taught that they should not smoke.

Things have changed greatly since those good days; and now young and old, boys and men, all smoke. They seem to think that the new ways of the whiteman are right, but I do not. In olden days we Hidatsas took good care of our bodies, as is not done now.

The tobacco garden

The old men of my tribe who smoked had each a tobacco garden planted not very far away from our corn fields, but never in the same plot with one. Two of these tobacco gardens were near the village up on the top of some rising ground; they were owned by two old men: Bad Horn and Bear Looks Up. The earthlodges of these old men stood a little way out of the village, and their tobacco gardens were not far away. Bear Looks Up called my father "brother," and I often visited his lodge.

Tobacco gardens, as I remember them, were almost universal in my tribe when I was five or six years of age. They were still commonly planted when I was twelve years old; but whitemen had been bringing in their tobacco and selling it at the traders' stores for some years, and our tobacco gardens were becoming neglected.

As late as when I was sixteen my father still kept his tobacco garden, but since that day individual gardens have not been kept in my tribe. Instead, just a little space in the vegetable garden is planted with seed if the owner wishes to raise tobacco.

The seed we use is the same that we planted in old times. A big insect that we call the "tobacco blower" used always to be found around our tobacco gardens, and this insect still appears about the little patches of tobacco that we plant.

The reason that tobacco gardens were planted apart from our vegetable fields in old times was that the tobacco plants have a strong smell which affects the corn. If tobacco is planted near the corn, the growing corn stalks turn yellow and the corn is not so good. Tobacco plants were therefore kept out of our corn fields. We do not follow this custom now, and I do not think our new way is as good for the corn.

Planting

Tobacco seed was planted at the same time sunflower seed was planted [late April].

The owner took a hoe and made soft every foot of the tobacco garden, and with a rake he made the loosened soil level and smooth.

He marked the ground with a stick into rows about eighteen inches apart. He opened a little package of seed, poured the seed into his left palm, and with his right sowed the seed very thickly in the row. He covered the newly sowed seed very lightly with soil which he raked with his hand.

When rain came, and warmth, the seeds sprouted. The seed having been planted thickly, the plants came up thickly, so that they had to be thinned out. The owner of the garden would weed out the weak plants, leaving only the stronger standing.

The earth about each plant was hilled up about it with a buffalo rib, into a little hill like a corn hill. It was a common thing to see an old man working in his tobacco garden with one of these ribs. Young men seldom worked in the tobacco gardens; not using tobacco very much, they cared little about it.

Arrowhead Earring's tobacco garden

An old man I remember, named Arrowhead Earring, had a patch of tobacco along the edge of a field on the east side of the village. He was a very old man. He used a big buffalo rib, sharpened on the edge, to work the soil and cultivate his tobacco. He caught the rib in his hands by both ends with the edge downward; and stooping over, he scraped the soil toward him, now and then raising the rib up and loosening the earth with the point at one end—poking up the soil, so to speak.

He wore no shirt as he worked, but he had a buffalo robe about his middle on which he knelt as he worked.

Small Ankle's cultivation

My father always attended to the planting of his tobacco garden. When the seed sprouted he thinned out the plants, weeded the ground and hilled up the tobacco plants later with his own hands.

Tobacco plants often came up wild from seed dropped by the cultivated plants. These wild plants seemed just as good as the cultivated ones. There seemed little preference between them.

Harvesting the blossoms

Tobacco plants began to blossom about the middle of June and picking then began. Tobacco was gathered in two harvests. The first harvest was of the blossoms, which we reckoned the best part of the plant for smoking. Old men were fond of smoking them.

Blossoms were picked regularly every fourth day after the season set in. If we neglected to pick them until the fifth day, the blossoms would begin to seed.

This picking of the blossoms my father often did; but as he was old, and the work was slow and took a long time, my sister and I used to help him.

I well remember how my sister and I used to go out in late summer when the plants were in bloom and gather the white blossoms. These I would pluck from the plants, pinching them off with my thumbnail. Picking blossoms was tedious work. The tobacco got into one's eyes and made them smart just as whitemen's onions do today.

We picked, as I have said, every fourth day. Only the green part of the blossom was kept [the calyx]. The white part I always threw away; it was of no value.

To receive the blossoms I took a small basket with me to the garden. There were two kinds used: one was the bark basket that we wove, and the other kind was made of a buffalo bull's scrotum with hair-side out.

Such a basket as the latter was a little larger than the crown of a whiteman's hat—the hat band being about the same diameter as the rim that we put on the basket. It had the usual band to go over forehead or shoulders. I bore the basket in the usual way on my back; or I could swing it around on my breast when actually picking, thus making it easy to drop the blossoms into it.

More often, however, I took the basket off and set it on the ground when plucking blossoms. I would make a little round place in the soft soil with my hands and set the basket in it, so that it would stand upright. The basket did not collapse, for the skin covering was tough and rigid, not soft.

I often used the scrotum basket also for picking chokecherries or juneberries. It was more convenient when berrying to carry the basket swung around on my breast. Going home with the basket filled with berries, I bore it in the usual way on my back.

My father usually worked with us; and indeed it was to help him, because he was old, that we picked the blossoms at all. It was slow work. I did not expect to gather more than a fourth of a small basketful every four days; and, as the blossoms shrunk a good deal in drying, a day's picking looked rather scant.

When we fetched the blossoms home to the lodge, my father would spread a dry hide on the floor in front of his sacred objects of the Big Bird's ceremony; they were two skulls [bird] and a sacred pipe, wrapped in a bundle and lying on a kind of stand. We regarded these objects as a kind of shrine. Nobody ever walked between the fire and the shrine as that would have been a kind of disrespect to the gods. My father spread the new-plucked blossoms on the hide to dry. Lying here before the shrine, it was certain no one would forget and step on the blossoms.

It took quite a time to dry the blossoms. If the weather was damp and murky for several days, my father, on the appearance of the sun again, would move the hide over to a place where the sun shining through the smoke hole would fall on the blossoms. The smoke hole, being rather large, would let through quite a strong sunbeam, and the drying blossoms were kept directly in the beam.

When the blossoms had quite dried, my father fetched them over near the fireplace and put them on a small skin or on a plank. We commonly had planks or boards split from cottonwood trunks lying in the lodge. They had many uses.

My father then took a piece of buffalo fat, thrust it on the end of a stick and roasted it slowly over the coals. This piece of hot fat he touched lightly here and there to the piled-up blossoms so as to oil them slightly but not too much. He next moved the skin or board down over the edge of the fire pit, tipping it slightly so that the heat from the fire would strike the blossoms. Here he left them a little while but watching them all the time. Now and then he would gently stir the pile of blossoms with a little stick so that the whole mass might be oiled equally.

This done my father took up the blossoms and put them into his tobacco bag. The tobacco bag that we used then was exactly like that used to-day, ornamented with quills or bead work; only in those days old men never bothered to ornament their tobacco bags, just having them plain.

When my father wanted to smoke these dried blossoms, he drew them from his tobacco bag and chopped them fine with a knife, a pipeful at a time. Cured in this way, tobacco blossoms were called *aduataakidu'cki*. They were smoked unmixed by old men.

The blossoms were always dried within the lodge. If dried without, the sun and air took away their strength.

Harvesting the plants

About harvest time, just before frost came, the rest of the plants were gathered—the stems and leaves, I mean, left after the harvesting of the blossoms. My father attended to this. He took no basket but fetched the plants in his arms.

He dried the plants in the lodge near the place where the cache pit lay. For this he took sticks, about fifteen inches long, and thrust them over the beam between two of the exterior supporting posts so that the sticks pointed a little upwards. On each of these sticks he hung two or three tobacco plants by thrusting the plants, root up, upon the stick but without tying them.

When dry, these plants were taken down and put into a bag or a package made by folding a piece of old tent cover over them. The package or bag was stored away in the cache pit.

When the tobacco plants were quite dry, the leaves readily fell off. Leaves that remained on the plants were smoked, of course, but it was the stems that furnished most of the smoking. They were treated with buffalo fat like the blossoms before being put into the tobacco pouch. We did not treat tobacco with buffalo fat except as needed for use and to be put into the tobacco pouch ready for smoking.

I do not remember that my father ever saved any of the blossoms to store away in the cache pit as he did the stem or plant tobacco. Friends and visitors were always coming and going; and when they came into the lodge my father would smoke with them, using the blossoms first, because they were his best tobacco. In this way, the blossoms were used up about as fast as they were gathered.

Before putting the tobacco away in the cache pit, my father was careful to put aside seed for the next year's planting. He gathered the black seeds into a small bundle about as big as my fingers hunched together, or about the size of a baby's fist, wrapping them up in a piece of soft skin which he tied with a string. He made two or three of these bundles and tied them to the top of his bed, or to a post nearby, where there was no danger of their being disturbed.

We had no way of selecting tobacco seed. We just gathered any seed that was borne on the plants. Of course there were always good and bad seeds in every package; but as the owner of a tobacco garden always planted his seed very thickly he was able to weed out all the weak plants as they came up, as I have already explained.

A tobacco plant, pulled up and hung up in the lodge, we called o'puti: opi, tobacco, and uti, base, foundation, substantial part.

The Mandans and Arikaras raised tobacco exactly as we did, in little gardens.

Selling to the Sioux

We used to sell a good deal of tobacco to the Sioux. They called it Pana'nitachani, or Ree's tobacco. A bunch six or seven inches in diameter, bound together, we sold for one tanned hide.

Size of tobacco garden

My father's tobacco garden, when I was a little girl, was somewhat larger than this room; and that, as you measure it, is twenty-one by eighteen feet. I have seen other tobacco gardens planted by old men that measured somewhat larger, but this was about the average size.

Customs

If any one went into a tobacco garden and took tobacco without notifying the owner we said that his hair would fall out; and if any one in the village began to lose his hair, and it kept coming out when he brushed it, we would laugh and say, "Hey, hey, you man! You have been stealing tobacco!"

"What? You say you got this tobacco out of Wolf Chief's garden without asking?" (laughing heartily) "Then be sure your hair will fall out when you comb it. Just watch, and see if it doesn't!"

I have said that my father softened the soil of his tobacco garden with a hoe. After the plants began to grow, the hoe was not used, either for cutting the weeds or for hilling up the plants. I have said that the weak plants were culled out by hand and that the strong plants were hilled up with a buffalo rib.

Buffalobird-woman (vol. 20, 1916: 154)

Note on tobacco

In old times we were careful about picking time, and never picked tobacco after seeds were formed.

The Sioux Indians had no tobacco. They got it of us, and they came every year to buy—which they were fairly crazy to do.

Tobacco 9b

HIDATSA NAME: *Opi* (Tobacco)
LOCAL ENGLISH NAME: Tobacco
BOTANICAL NAME: *Nicotiana quadrivalvis* Pursh var. *bigelovii* (Torr.)
DeWolf

"Of the Tobacco Plant" is Wilson's edited version of his field notes recording an interview with Wolf Chief that he sent to Clark Wissler. I have done a little more editing to provide context, continuity, and clarity.

Wolf Chief (vol. 17, 1915: 65–78)

Of the Tobacco Plant

Wolf Chief had picked a quantity of tobacco blossoms and brought them to the lodge on August 2. He laid them in the bottom of a shallow wooden box which he leaned against the side of his cabin so that the sun would fall on the blossoms and dry them. I photographed the blossoms, and Good Bird drew a sketch of two or three freshly picked blossoms.

Discussing these tobacco blossoms, Wolf Chief said: "My father's family used to plant this tobacco. We covered the seeds very thinly with earth. My father said also, 'We Hidatsas used to throw a little wood ashes on the soil of our tobacco gardens for then the plants grew better. Also, if we were able, we tried to get ground that was in some low place or spot where it would always be moist and damp. In this soil we planted the tobacco seeds, quite early in the season.'

"I think the reason that the tobacco seed was planted early was in order that the blossoms would come early in the summer. They could then be dried and smoked, for we Hidatsa considered the blossoms to be the best part of the plant.

"When the plants grew up they were picked free of weeds, but

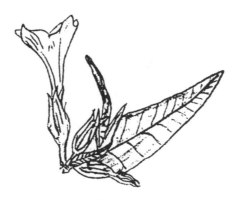

22. A tobacco blossom with the calyx prized for smoking (nearly life size).
Drawing by Goodbird. (Courtesy of the Minnesota Historical Society)

the ground was not hilled up about them as was done with our corn plants. The owner of the tobacco garden watched for the blossoms and when they came he picked them every four days. If, when he went to pick the blossoms, he discovered that some had been missed by him at the last picking, he just left them on the plant to produce seed.

I remember that my father always chose a spot where the best ground was for his tobacco patch. The tobacco crop was the very last to be gathered in the fall.

The tobacco harvest was in about the middle of September when the plants were dead and, in a measure, dry. My father used to gather up the plants and take them into his lodge to dry for many of the stalks would still be green and moist and need drying.

I remember that he used to take them to his lodge about four plants in each hand, grasped by the roots.

I have also seen some Hidatsa people load their tobacco plants into a blanket. A man and his wife would then bear the load to their lodge by holding the corners of the blanket. My father used to carry the plants quite carefully as he did not want to shake out any of the seeds. These plants he used to dry on poles stretched out on two of the outer supporting posts of his lodge.

There were two of these poles, and they were bound at either

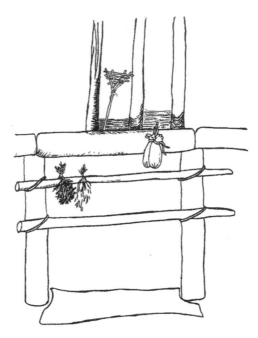

23. Drying and storing tobacco. The hide on the floor is meant to catch falling seeds. Drawing by Goodbird. (Courtesy of the Minnesota Historical Society)

end to the two posts, the tie being made by lapping a lariat around the post and the ends of the poles. The plants were straddled over the poles, roots up, or two of them might be bound together by their roots and strung over the poles, one plant on either side. The drawing (Figure 23) depicts this [on the left is a bundle of the lower or root half of some tobacco plants and on the right is a bundle of the top halves]. Of course the two poles were covered from end to end with the tobacco plants.

These poles were usually tent poles that were bound to the posts immediately back of the horse corral and in front of the post toward the fire. The tobacco was left here to dry on the poles until about November or for nearly two months.

My father then prepared to make it ready for winter. On the floor, directly under the drying plants, he spread an old tent-skin or some other skin as shown above.

My father now took down one of the tobacco plants. With his thumb and finger he squeezed the seed balls on the plant, letting the seeds fall on the skin. This he did to all the seed balls on each plant until every seed was out. These seeds made a package about as big as a man's fist.

The seeds he folded up in a bit of tent-skin into a package, tied it with a small thong, and put it in the bag in which he kept his arrow tools. This bag was made of tent-skin and was about two and a half feet by one foot, and it was tied with a thong that went around the top under a fold of the edge. When this thong was pulled, it drew the mouth of the bag tight. [Tent-skin is leather from an old tipi and had been thoroughly smoked by many fires and rolled and unrolled many times. It was soft and used for both carrying bags and moccasin tops.]

This bag was often hung on one of the outer posts near the Big Bird's shrine because here it was in no danger of being disturbed by anyone, for the ground around the Big Bird's sacred objects was held sacred.

The seeds, as they fell from the crushed seed balls, were scattered over the tent-skin and my father lifted up the edge of the tent-skin and made the seeds roll to the center where they were easily gathered up and put in a package.

The seeds disposed of, my father now picked up a plant and, grasping it in his two hands, he twisted and broke it through the middle. The plant was now divided into two halves, the top or more tender half and the heavy root and stem half. These two halves of the plant he laid in separate piles at either end of the skin, and he treated all the plants thusly.

These top parts were soft, and they were ready to be smoked at once; but the harder parts of the roots and stems needed pounding. For this he took a stone convenient to hold in the hand but rather flat, and with the edge he crushed the roots and stems. As he pounded he would stir up the pile now and then. He would work at one edge of the pile until the roots were well pounded, and then he would move the pounded part aside and begin to pound again on the edge of the pile yet remaining. The roots and stems were thus all pounded. My father sat as he did this.

This pounded part we called "root tobacco." When we wanted to smoke it we oiled it. A little lump of buffalo fat was taken in the palm, a little handful of pounded tobacco was laid upon it, and it was oiled by being squeezed around and over the lump again and again. When well oiled it was put down and another hand-ful was taken and squeezed upon the lump. The oiled tobacco was laid on a board in front of the fire near the fireplace where it would be exposed to the heat.

This oiling of the tobacco was done just as one wanted to smoke it and after it had been stored away. It was not oiled before storing.

If there was fat in the house this root tobacco was always oiled. The top tobacco, made from the more tender parts of the plants, was sometimes oiled and sometimes was not as taste or conve-nience might suggest. The root tobacco, after being oiled, was exposed before the fire until all the grease upon it was melted and it became shiny. Then the tobacco was taken from the fire and rubbed well in the hands. It was then again exposed to the fire until it lost its oily appearance and looked dry. The smoker then chopped it up with his knife, and it was ready for his pipe.

Trading tobacco and garden produce

Visitors from other tribes used to go nearly crazy about this tobacco. We used to sell them a double handful of our tobacco for one buffalo hide, sometimes a dried hide, sometimes a tanned one; or a double handful of the tobacco we sold for a package of dried meat three feet long and one foot thick. Women from other tribes also bought quantities of beans and half-boiled dry green corn. A wooden bowl containing about 400 cubic inches (figured out from measurements given by Wolf Chief in pantomime) of either beans or dried green corn brought one hide or a package of meat, and the same price was paid for one string of braided corn or for a string of squashes a foot in diameter and a little more than three feet long. Visiting Indians used to be almost crazy about these vegetables for they did not raise any themselves.

(To obtain an idea of quantity of beans I had Wolf Chief measure out what would fill a wooden bowl as described and found it came to three and a half pounds (about 1.6 kg). A string of braided corn should be about fifty-six large ears. [Wilson thought that this was about a bushel of corn, but a full-sized ear of Hidatsa corn grown in our garden weighs about 100 grams; therefore, fifty-six ears would weigh 5.6 kg—somewhat less than a fifth of a bushel of unshelled corn, which weighs 70 pounds or about 32 kg]. The string of squash "One foot in diameter and three feet long" is doubtless a string of dried squash, doubled into a package that size. It is evident the Hidatsas asked a pretty stiff price for their vegetable products as indeed they do yet. Presumably the visiting Indians purchased tobacco and vegetables to take home. Perhaps they were better content to pay a stiff price after they had been feasted by their hosts. GLW)

My mother thus one year bought one hundred hides with vegetables she had raised, and these hides she traded to a whiteman for a fine spotted horse, swift enough to chase buffalo.

Another year she sold vegetables to one family of Unkpapa Sioux [Hunkpapa] who gave her in exchange a middle-sized buckskin colored horse. She gave them vegetables of all kinds: corn, squash, and beans, but no sunflower seeds. I do not remember how much there was of what she gave them.

The use of tobacco blossoms

My father, as I have said, used to gather the blossoms from his tobacco plants every four days, drying them on a rawhide skin either outside the lodge or within it in a sunbeam. There was always a spot in the lodge where the sun fell on the floor through the smoke hole, and this we called the "sunbeam."

I think his custom was that if it was a sunny day, and there was no wind, he would lay the blossoms on the ground on the sunny side of the entranceway of the lodge or on the flat roof; but if the day was windy, he brought the blossoms inside the lodge and laid them in the sunbeam, moving them now and then to keep them in the sunbeam.

Blossoms dried in about two days. They were stored in a small skin sack a little larger than a tobacco pouch.

My father usually began smoking the blossoms as soon as they were dried, for they were thought the best part of the plant and smokers awaited the blossom season with much eagerness. My father used to smoke the blossoms mixed with kinnikinnick. He also stored away blossoms for winter smoking. I remember that he used to put away two sacks full of these dried blossoms which he hung above the beam that rested on top of the outer posts. The manner in which they were hung is shown in the drawing above. I think my father never sold any of these blossoms to any one but saved them to be used perhaps rather sparingly by himself and his friends, but he sold tobacco of the two other kinds—the root tobacco and the tender top tobacco. Most of the old men of our tribe raised tobacco, although some did not. I remember that some who did not would sometimes send a little boy to our lodge and ask my father for blossoms of the tobacco, but I remember also that he answered such a request with a very small quantity.

Whiteman's tobacco

The first tobacco brought to us by whitemen was chewing tobacco. My father tried to smoke it.

"I do not like this kind," he said, "it makes my head ache."

This chewing tobacco was issued to us by the government.

I once saw a long plug of the chewing tobacco offered to a buffalo skull on top of an earthlodge roof.

Top tobacco and root tobacco

The top tobacco and root tobacco were also stored away in sacks, each sack about two feet long and eighteen inches in diameter when filled. Such a sack was made by folding over a skin square and sewing the bottom on one side. The top was pursed and tied shut. Small Ankle, as I recollect, used to put up one sack of top tobacco and one sack of root tobacco of the size just described.

The use of kinnikinnick

[Kinnikinnick is a rather general term for certain plants which are mixed with tobacco, either to extend the actual amount of tobacco being smoked or to make it more pleasant to smoke. Usually it refers to either the inner bark of the red-osier dogwood (*Cornus sericea*) or the leaves of the bearberry (*Arctostaphylos uva-ursi*), but other plants may also be blended with tobacco.]

I have said that blossom tobacco was usually mixed with willow kinnikinnick [red-osier dogwood] when smoked. Root tobacco and top tobacco were usually smoked straight. We did not store kinnikinnick but went out and cut sticks and dried the bark whenever we wanted to smoke it.

When kinnikinnick was cut, the red outer rind of bark was shaved off the sticks and thrown into the fire as useless. The yellow bark lying just under the outer rind was also scraped off and thrown away. This left the white or inner bark still on the stick. The red and yellow parts of the bark were bitter. Red Stone I think was the most skillful kinnikinnick bark cutter that I ever knew.

The white or inner bark of the willow stick was now shaved off the green stick and laid on a drying frame to be dried.

This frame was held over the fire and the bark, loosely doubled up or folded, was soon dry. The frame when not in use was thrust over the beam on which hung the tobacco bags. It was made of a kinnikinnick rod five or six feet long and split into three and sometimes four tines. Small willow rods were woven into these tines to spread them apart. As shown in the drawing (Figure 23) it was thrust by the handle back over the beam and under the rafters. When we were in camp we often made smaller kinnikinnick drying frames bound with bark. That one which we kept in our lodge was also bound with bark but was larger and more carefully made. We used to make a kinnikinnick drying frame from a kinnikinnick rod because it would split and bend easily.

More about planting

My father used to tell me that the tobacco seed when planted should be covered but with a very little earth. I have found, however, that even when I plow under the ground in which tobacco is planted that it comes up the next year in my garden; and this it does every year, even when I plow quite deeply. I always raise tobacco every year here in my garden, but I seldom plant any because it comes up this way.

[Tobacco seed that falls on the ground (and each plant produces thousands of tiny seeds) is one of the first plants to sprout in the spring so that the gardener needs to thin the mass of sprouts to about a foot apart for best results. I find that "hilling" at least some plants is often necessary because they do blow over. Buffalobird-woman noted that it was hilled as well.]

My father's tobacco patch was about nine yards long and about six yards wide as I here pace it on the ground. He prepared the ground for his seed, first raking the soil well with an ash-wood rake. He scraped a shallow furrow up in the soil in which he sowed his seed. He covered the seed very lightly by raking the soil with his rake. Three feet over he made another furrow and this he also planted with seed. Thus his tobacco grew up in long rows three feet apart.

To make the furrow he held the rake in both hands with the point of the handle resting on the ground and dragged it after him as he walked backwards or sidewise pulling it.

His tobacco patch was in old corn ground, and he used no hoe or digging-stick in its preparation. I have said that the rows were about three feet apart. When the plants appeared he carefully pulled out any weeds that came up with them, but as far as I know he did not thin them out. [He definitely had to thin them out because the plants come up by the hundreds and will choke one another out.]

Some of the old men in our village I think sowed tobacco seed broadcast, but I am not very sure of this. I feel sure, however, that plants sowed in rows grew up stronger and taller. [This is not my experience at all, and I have grown it for about twenty years.]

Although hoes were not used in preparing the soil they were used to cut down the weeds that came up, but my father used to pull out the weeds with his hands from among the plants first. Then he hoed between them. He hoed in the summer season. A space of about ten feet separated his tobacco patch from the adjoining corn field as the strong smell of the tobacco would soak into the corn and make it taste bad.

Wood ashes were very good to put on tobacco soil, and I have often heard my father tell me that ashes from the lodge fireplace put on a tobacco patch made the plants grow stronger and taller. I do not know where my father learned this, and I never saw him do it myself. My father used the same patch for his tobacco each year. If the season was a dry one, the tobacco made a poor crop, but if it was a damp season and there was plenty of rain, the plants grew strong and the crop was abundant.

[Again from my personal experience it turns out that the issue is how dry is dry and how wet is wet? This tobacco does not like to be wet for long nor will it grow in very dry soil. It prefers a fairly light soil (sandy loam) and regular rains, although not to the point of saturation. Thus growing tobacco in west-central North Dakota is even more uncertain than growing maize. The tobacco is also blown down by the winds that typically accompany thunderstorms; then comes the tedious and smelly job of setting the plants upright and mounding some wet soil around the base of most plants to keep them from tipping. I have found that three years is about the maximum number of times a piece of ground can be used for tobacco. After that it grows very poorly.]

My father, when he sowed the rows in his tobacco patch, sowed from the little seed bundle which he had put up the autumn before.

Wolf Chief (vol. 18, 1915: 440)

Note on smoking tobacco

This that I now give you in this pipe to smoke is dried blossoms [calyxes] of native tobacco. Yesterday when you tried to smoke, it seemed to choke you. This will not do so today because I oiled the dry blossoms with kidney fat of a steer.

Our object in oiling our tobacco was to make it mild and not harsh to smoke.

Wolf Chief (vol. 30, 1915: 147)

Small Ankle's advice against smoking

My father warned me when I was young not to smoke tobacco. "Some day," he said, "enemies will pursue you against the sunrise, and your mouth will become filled with thick saliva that will fill your air passages if you smoke. Therefore do not smoke. Not to do so will someday save your life."

My father knew it was hard to run against the sunrise. One did not then feel strong and boys who were smokers became short winded. It was hard to run against the sun because of the heat especially if one was a smoker. One could run very much better if the sun was behind him.

The origins of tobacco

Butterfly is relating the story of the origins of the Hidatsa and their world. This was the work of two gods—Itsikamahidish or First Worker and One Man. Upon completion they were showing each other what each had accomplished—One Man created the world south of the Missouri River, and First Worker created the world north of the river. There was a certain amount of competition involved.

Butterfly (vol. 9, 1910: 85–88)

They came to the south side, and First Worker showed One Man what he had done. One Man liked the white buffaloes, elks and the people. All these the gods had created by the mind. I also think that First Worker made all the tribes, but I do not know the story. I can only tell the story as I heard it.

But before One Man and First Worker made the grass and trees they went all over the world and found the ground to be just sand. There was as yet no grass or living plants upon it, but they found little tracks. "What is this?" they cried. They followed the tracks and overtook a little mouse. "*How*, my friend," said the two gods.

"Not so," said the mouse—a female mouse, "you are wrong. You are not my friends. You are my grandsons."

"You must be our grandmother," they said.

"Yes," she answered.

They went on and found other tracks. These they followed and came to a large female toad. "*How*, friend," they said.

In answer the toad said, "I am an old woman, and you are my grandsons."

"Where did you come from?"

"This ground is my body, and for that reason I stay here on top," the toad answered.

They went on and came to a great buffalo bull lying on a hill. First Worker went on in a different direction, but One Man came to the buffalo bull and said, "*How*, friend!"

"You are right," said the bull, "I am your friend!"

"I have a pipe," said One Man, "but I have nothing to smoke in it."

"I will see that you have some tobacco," said the bull, "some *Opi*!" About the middle of summer when you hear a noise, run hither."

"I will do so," said One Man, turning to go.

As One Man departed, the buffalo bull dropped his urine. It fell on the ground like tobacco seeds, and tobacco plants sprang up.

About midsummer One Man heard a great noise like "hum-m-m!" It was made by a kind of a bee or fly, an inch long, that we call tobacco-blowing bee. One Man went in the direction of the sound and found plants with white blossoms and many of these bees or flies flying about them as they do about tobacco plants.

One Man plucked the white blossoms of the plants but said, "I have nothing to dry them on."

"Take my hair," said the buffalo bull and gave him some of his shed hair, for all buffaloes shed their hair like felt. One Man dried the blossoms and then said, "I have no fire!"

"Go on the other side of the hill, and you will find a man there named Ear Afire." One Man went as he had been told and found Ear Afire.

Ear Afire taught One Man how to make a fire when he wanted to smoke. One Man lighted his pipe and gave it to Ear Afire to smoke. He then returned to the buffalo bull and gave him the pipe to smoke also.

This Ear Afire who taught One Man to make a fire and smoke represented burning ground, for there are places about this reservation where the ground is afire, and we can see the smoke.

(Butterfly here refers to burning coal deposits [layers of lignite—a form of soft coal] such as are formed in the badland region west of the Missouri. At least one of these burning mines is found on the Ft. Berthold Reservation. GLW)

One Man said, "I will make people resembling me, and I will give them ceremonies; and you," he said to the buffalo bull, "shall stay with them and shall be their leader." For this reason in every ceremony we must have a buffalo skull.

(Butterfly does not say who these people are. One Man is the patron of Mandans, First Worker of the Hidatsas in the mythology of these tribes. GLW)

The first pipe and the first tobacco

This story was probably told to Wilson by Goodbird during Wilson's first year at Fort Berthold. (vol. 3, 1906: 38–40):

Numakmaxina (Only Man) wanted to smoke. He knew he ought to smoke. So his Hidatsa people made him a pipe of hardwood. And then they made a stem. The pipe was made of ash. The pipe is called male or man, the stem is called woman or female, and where they were joined was the Missouri. They went around for tobacco. They asked the buffalo.

"Yes, I can make it." He rolled over, got up, pissed, and said, "When you hear a big noise, come at once. Hurry."

They went away. In the summertime when they heard, "Boo-oo-oo-oo," they came. The sound was the tobacco fly—bees. They heard the sound and came.

"Pick the white blossoms. They are best. Also, all the other is good."

"But what shall I put it in?"

Down went the buffalo and rolled, and when he got up his hair came off. "Put the tobacco leaves on this. They will get dry."

"But where will we get fire?"

"Of this man," and he took him to the Fire Man.

"Look," said Numakmaxina, "I have tobacco in my pipe but no fire. Let us smoke together."

The Fire Man took fire out of his ears and put it in the pipe and began to smoke.

Numakmaxina wanted to take some along. The Fire Man took some out of his ears, and Numakmaxina took the fire with him.

Nowadays, whenever they dry tobacco, they find buffalo hairs in the tobacco.

FAMILY *Solanaceae*–potato family
GENUS *Nicotiana* L.–tobacco
SPECIES *Nicotiana quadrivalvis* Pursh–Indian tobacco
VARIETY *Nicotiana quadrivalvis* Pursh var. *bigelovii* (Torr.)
DeWolf—Bigelow's tobacco

Red-osier dogwood (smoking, sweet berries)

HIDATSA NAME: *opi ihaca* (tobacco mixed or mixture)
LOCAL ENGLISH NAME: kinnikinnick (red-osier dogwood)
BOTANICAL NAME: (*Cornus sericea* L.)

Kinnikinnick is a name used for several plants or mixtures of plants. Red-osier is one of the more common sources, but bearberries (*Arctostaphylos uva-ursi*) are also called kinnikinnick, as is the inner bark of some willows. It is basically a "tobacco helper" to make valuable tobacco last longer, and it also makes the tobacco less harsh as it is smoked. Red-osier is often found along waterways in rich soils, and the numerous and very conspicuous white berries are quickly consumed by various species of birds (and, after frost, humans).

Buffalobird-woman (vol. 20, 1916: 266)

The inner bark of kinnikinnick we used for smoking, but only in recent years after meeting the whitemen. In old days we did not use it. When I was a little girl our old men smoked native tobacco and did not mix it with kinnikinnick.

My father smoked very frequently, but he never mixed his tobacco with kinnikinnick until he got to using whiteman's tobacco. He never mixed anything but fats with his tobacco.

Kinnikinnick sticks make pretty good firewood in the absence of anything better, but they were rather light.

The kinnikinnick bush bears a white berry. These berries are quite bitter before frost falls, but after freezing they are quite sweet. We used to eat them even when bitter, picking them from the tree. After they are frozen they are quite good for two or three days, after which they turn black and fall on the ground. We were fond of the sweet, new frozen berries before they fell off the bush.

I would gather one or two gallons of berries in a bark basket, take it to the river, and sink it gently until half under water. The bark, bits of leaves, and sticks, etc., would rise and could be taken off while the berries sank to the bottom of the basket. The brief wetting did not hurt the basket. The berries were taken to the lodge to be distributed to all who would partake. We gathered berries after the frost, or even before if they were well ripened.

FAMILY *Cornaceae*–dogwood family
GENUS *Cornus* L.–dogwood
SPECIES *Cornus sericea* L.–red-osier dogwood
SUBSPECIES *Cornus sericea* L. ssp. *sericea*–red-osier dogwood

Bearberry (edible berries)

HIDATSA NAME: *ama-matsu*
LOCAL ENGLISH NAME: ground-cherry (bearberry or kinnikinnick)
BOTANICAL NAME: *Arctostaphylos uva-ursi* (L.) Spreng.

Although Wilson identifies this plant as a ground-cherry (or Good-bird translates his mother's words as such), the description of it by Buffalobird-woman does not fit that identification in any way. In the first place the fruits of the two species of ground-cherry (*Physalis virginiana* and *P. heterophylla*) that grow on or in the vicinity of the reservation are yellow or orange and surrounded by papery husks. Ground-cherries are fairly common plants in midgrass and tallgrass prairies, and it is surprising that Buffalobird-woman does not mention them, because the fruits are both edible and sweet. Bearberries are red and, as Buffalobird-woman notes, the plants are found on the tops of ridges in "stony places." Buffalobird-woman speaks of eating them fresh from the vine, which poses a problem (probably a translation problem) because neither ground-cherries nor bearberries are vines. The bearberry plant is a low, sprawling shrub that Wolf Chief, in his very different account of this plant, also describes as a vine. People of many tribes still collect the bearberry as a source of kinnikinnick. Perhaps because Hidatsa women did not smoke, Buffalobird-woman did not think of that aspect of the plant as being significant, but Wolf Chief certainly did.

Buffalobird-woman (vol. 20, 1916: 327)

Ground-cherries grow on the tops of the high hills or buttes on this reservation and elsewhere in North Dakota. They grow in stony places and taste much like chokecherries.

When ripe they are dark red and a little larger than chokecherries. This plant grows about a foot high.

When we found a patch of the plants with ripe berries and went out to gather them, we were always careful to approach them from downwind. Thus if the wind was blowing from the west, we approached from the east; if from the east, we approached from the west. If we did not do so, the wind would carry our smell to the fruit and the cherries would turn sour. Approaching from the downwind side, we found the cherries were always sweet. Everyone on the reservation knew this—that if the picker approached going with the wind he would find the fruit spoiled and sour.

Ground-cherries are scarce with not many being found usually. If a large quantity were found, however, they were sometimes pounded up and made into lumps like chokecherries. They were eaten fresh either from the vine or brought home. They were gathered in rawhide baskets or regular berry baskets and when fetched were passed around to all in the lodge.

FAMILY *Ericaceae*–heath family
GENUS *Arctostaphylos* Adans.–manzanita
SPECIES *Arctostaphylos uva-ursi* (L.) Spreng.–kinnikinnick

Bearberry or kinnikinnick (smoking)

HIDATSA NAME: *Opi ihaca kati* (tobacco mixed real =
real tobacco mixture)
LOCAL ENGLISH NAME: (none given) (bearberry or kinnikinnick)
BOTANICAL NAME: *Arctostaphylos uva-ursi* (L.) Spreng.

Wolf Chief said that "men gathered this plant" both to extend
their tobacco supply and to make "whiteman's tobacco" easier to
smoke. Buffalobird-woman approached the plant from the perspective of a berry-picker. She had one name for the plant, and
Wolf Chief had another. Wolf Chief even noted that the berries
were blue (he was probably thinking of the blue juniper "berries,"
which are cones), but this is more than likely a lapse of memory,
and the berries were not of interest to him at all. Note that the
word *kinnikinnick* can mean the bearberry, the inner bark of red-
osier dogwood, and the inner bark of a willow. At present there
are any number of mixtures of shredded bark, inner bark, and/
or crushed leaves that are called kinnikinnick.

It might be worth considering that the plants which are now
additives were smoked before the acquisition of tobacco. The
name "real tobacco mixture" almost implies that this was the
case. Tobacco seeds are tiny and found only infrequently during
archaeological excavations.

Wolf Chief (vol. 20, 1916: 328–31)

This plant grows on the sides of bluffs and hills. It is not found in
many places west of the Missouri, but there is, I hear, one place
on the Little Knife River where it grows. In this part of the reservation it grows in only two places—one place nine the other
six miles from Independence. Also near Blue Butte there is quite
a good deal. This is about twenty-five miles from Independence
on the bluffs on the Little Missouri.

Killdeer Mountain is nearly covered with this kind of plant. There are two hills together, one rather round and the other about two miles long. The round hill is called the "head," and the other is called in our language *bahic*. In old language *bahic* means taller or higher. But in this day *pahi* means to sing. I do not know whether the hill is named from *bahic* or *pahi*. But I think the former, because the hill is the tallest in the neighborhood.

The plant is about the length of my arm and trails on the ground, full of branches and green leaves. These leaves stay on all winter and keep green all the time. They bear little blue [*sic*] berries. These come on the vine every summer.

When gathering the plant I plucked it off the root—that is, broke it from the root in the ground, as we did not use the root. We would gather a pile of about a peck at a time and would wrap it in a bundle in my saddle blanket. When we got home we sometimes put the plants in a pot with some buffalo meat broth. This might be broth of dried or fresh meat. The pot was put on the fire and let boil a little while. Then the plants were taken out and hung on the drying pole over the fireplace or outside on the corn stage. Or, if we were in camp, we put three forked sticks together like a tipi and tied the plants in a bunch at the top. They dried in a day or two.

The plants were put in the broth because this made them sweeter to smoke.

When quite dry, I pulled the leaves off with my hand and put them in a little sack.

Cooking the leaves in the broth also makes them easier to break up with the fingers to mix with my tobacco. I would put some in my left palm and crumble them with the thumb and fingers of my right hand.

When I smoke it, I mix equal parts of the crumbled leaves and our native Hidatsa tobacco.

Men usually gathered this plant when out hunting game.

Finding the plant, the hunter would bring some home or into camp.

The plant could be gathered at any time of the year, and there was no seasonal gathering of it—so as to store some for winter, for example.

I have heard that this plant has been used since very old times. Also I have heard that in very old times the Hidatsas smoked tobacco straight, wasting nothing, but picked off the blossoms every four days and pulled up the roots and put away leaves and stems and roots. But as they smoked this with visitors, they were apt to run short toward the end of the season. Then they found that this plant, if mixed with tobacco, would stretch out their supply. Also it made smoking milder, and they liked that, too.

Whitemen's tobacco came, and we found it strong, so we began to mix it with other things—kinnikinnick bark, rose leaves, ground cherry leaves, buckbrush leaves, and leaves of another plant that turns red in the fall [probably fragrant sumac, *Rhus aromatica* Ait.]. We did this, I mean, on war parties. This mixing, however, is a recent custom. We never smoked any of these plants unmixed—always mixed with native or whiteman's tobacco.

But in old times, though we used *opi ihaca kati* mixed without tobacco, we did not use these other plants I spoke of. I am not sure, but I think that the use of kinnikinnick bark and rose leaves began about eighty years ago. This is the best information I have—but I can't speak with certainty.

Once I got a box of apples, and there were a few leaves in it. I mixed these apple leaves with my tobacco and found it smoked good. I wish you could send me some apple leaves.

Sometimes we got a kind of root from the Crow Indians that we used to mix with tobacco. It smelled good and opened our lungs. I do not know the plant it is from. I think they must get it from the Flat Head Indians, for in my young days I visited the Crows but never heard of the plant.

Goodbird (vol. 9, 1910: 349)

True tobacco mixture

Five miles west of Independence on the side of the hill is a small patch of a running, woody ground plant of rich green, bearing a red berry. We Hidatsa call it *opiiasakati* or tobacco mixed really. A free translation is "true tobacco mixture."

We Hidatsa mixed kinnikinnick or red willow bark [red-osier dogwood] and also the leaves of the wild rose with our tobacco. These leaves we dried by holding the bush or branch over a fire and crumpling the dried leaves in the hand. We also used the red leaves of a plant which grows somewhere around this reservation. I do not know the place where it grows, but others of the tribe do. [Fragrant sumac (*Rhus aromatica* Aiton) is common in the Badlands to the west.]

But this true mixture is what we prize most for smoking. This plant grows in a much larger bed out at a place about eight miles from here southwest of Skunk Creek. We usually go to this place on Skunk Creek for our supply of this plant to dry. We gather this plant at any time of the year, even in winter. We gather rose leaves at most seasons of the year, although we cannot do so in winter.

FAMILY *Ericaceae*–heath family
GENUS *Arctostaphylos* Adans.–manzanita
SPECIES *Arctostaphylos uva-ursi* (L.) Spreng.–kinnikinnick, bearberry

NINE

Plants Used for Dye and Coloring

Yellow owl's-clover (dye)

HIDATSA NAME: *i'xoka io'te* (kit fox dye)
LOCAL ENGLISH NAME: kit fox dye (yellow owl's-clover)
BOTANICAL NAME: *Orthocarpus luteus* Nutt.

Buffalobird-woman (vol. 20, 1916: 205)

This plant grows abundantly on the prairie on this reservation. It was used in old times for dying red, though I never used it myself. In my time we had come to use whitemen's dyes a good deal. We used to take whitemen's cloth and cut it up into pieces to make dye for porcupine quills. Thus we made green dye for quills from green blanket cloth.

My father, I remember, colored the horse hair tassels on his coat with kit fox dye.

The plant itself was used, not the root.

A piece of tent-skin about a foot or a foot and a half square was laid down. The plants were scattered over the skin rather thinly. Then they were lightly crushed with a stone. Over these lay porcupine quills, wetted. Over the quills again, macerated kit fox dye plants. The quills, as I have said, were wetted by dipping in cold water.

The skin was now folded over flat and laid under the skin of one's bed for a night or two when red spots appeared on the quills. Sprinkle a little more water over them, put back under bed-skin, and in a few days more the quills would have turned a bright red.

FAMILY *Scrophulariaceae*–figworts, scrofulaires
GENUS *Orthocarpus* Nutt.–orthocarpus, owl's-clover, owlclover
SPECIES *Orthocarpus luteus* Nutt.–yellow owl's-clover, yellow owlclover

Water smartweed (dye)

HIDATSA NAME: *Kadakadaduti*
LOCAL ENGLISH NAME: pink top (water smartweed
or water knotweed)
BOTANICAL NAME: *Polygonum amphibium* L.

This is a very common plant around sloughs and along sandy river banks.

Buffalobird-woman (vol. 20, 1916: 272)

This plant grows three or more feet high with joints like a reed, broad leaves, and pink blossoms. It grows in wet places.

We dug out a few roots and boiled them. In the dye so made I colored porcupine quills a light yellow. My father also colored white horse hair in this dye. The white horse hair was boiled in the dye. I never tried to dye gull quills in this dye. Perhaps others did. I do not know, but I never did.

However, the color was not too good, being too light.

The foregoing dye, and others which I give you this summer, are all the native dyes that I know. As I have already said, when I was young, we were already coming to depend chiefly upon whitemen's dyes for coloring quills.

FAMILY *Polygonaceae*–knotweed family
GENUS *Polygonum* L.–knotweed, smartweed species
SPECIES *Polygonum amphibium* L.–water knotweed, water smartweed

Dye plants—unidentified

LOCAL ENGLISH names: (none given)

Buffalobird-woman (vol. 20, 1916: 346)

I know of but three native yellow dyes, for in my day we had come to use whiteman's dyes a good deal. [She provides information on only two.]

Hidatsa name: Maicite

A yellow dye that comes from the Rocky Mountains in Montana. It grows on dead, rotten pines. Not on this reservation. (Probably—indeed certainly—a moss. GLW) [Most probably the lichen *Letharia vulpina* or wolf lichen, which was used by many groups of people as a source of yellow dye (Grinnel 1905, 43; Johnson 1970, 304; Moerman 2009).]

Hidatsa name: Mika maiote *(grass coloring)*

I think this grass probably grows around Independence, though I know of none myself. It was used to color gull quills. The plant grows mixed in with buckbrush in buckbrush patches. Grows about 2 feet high with a white flower.

The roots of the plant were boiled, and while the liquor was still hot, the gull quills—all prepared for use—were put in for a short time.

Meanwhile, the woman had taken ashes of dead elm bark—no other would do—and put the ashes in warm water. She stirred the ashes about, then let them settle, and poured the lye off into a pan or bowl or other vessel.

The gull quills were dipped into the dye and held till of the color desired. The quills were of gull wing feathers, in bunches of about ten each.

When the quills came out of the liquor made by boiling the roots, they were of a very light pinkish color. After passing through the ash water (lye), the quills became dark. Indeed, when finally taken out, they were of a dark or nearly black color but with some reddish or brownish-red showing. Inconspicuously

FAMILY *Parmeliaceae*
GENUS *Letharia* (Th. Fr.) Zahlbr.–wolf lichen
SPECIES *Letharia vulpine* (L.) Hue–wolf lichen

TEN

Plants Used for Toys

UMAKIXEKE, or game of throwing sticks

Buffalobird-woman and GLW (vol. 6, 1908: 131)

The name *umakixeke*, or at least the first syllable, Goodbird thinks means to strike with a glancing blow, and *ki* is a diminutive. The players take turns throwing a dry and peeled young willow or boxelder stick. In throwing, the forefinger is hooked over the smaller end and the sticks are thrown the greater end first. These are thrown against the ground (or a flat stone) and made to spring upward. The one whose stick travels farthest wins. The game started down in the five villages where, down near the Knife River, there was a big flat stone that was nearly worn through.

Children begin to play this game at the age of four, and it is much played by schoolchildren in the spring. Sticks are coveted which have a spring. Goodbird's derivation of the name is not very satisfactory. GLW

Goodbird (vol. 15, 1918: 421)

Note about Uwkiheke *(Umakixeke)*

When I was a boy, I played *umakixeke* all day, until my forefinger had a sore spot in the middle of the tip. Then I used the second finger.

We boys always played one against another.

In playing, the one who won was the boy whose stick went the farthest. We often said, "We will throw ten sticks," or "We will throw two sticks," or "We will throw one stick." Also we would say, "You throw first, and next time I will throw first."

The sticks were thrown against the ground and rebounded quite a distance. By long practice a boy knew just what kind of earth and what contour of ground best aided his stick to go farthest. One became quite expert in choosing, even in a small area, the exact spot most likely to fulfill these aims.

Popguns

Buffalo Bird Woman (vol. 20, 1916: 154)

Boys made popguns. They got young boxelder or young ash and burned out the pith with red hot wire. Popguns I have heard were used in my tribe in old times. They were in use when I was a little girl.

[Popguns were used for firing a projectile (the "wad"). Wads were made by chewing the inner bark of the elm. The rod was a juneberry shoot. The wad was inserted into the hollow stick and was ejected by rapidly shoving the rod into the stick, which compressed the air, which ejected the wad with a "pop."]

A toy horse

Buffalobird-woman (vol. 10, 1911: 194–95)

A kinnikinnick stick [red-osier dogwood (*Cornus sericea*) in this case] that had been thrown away was called a horse by playing children. A bit of hide was cut so that it would curl up at either end like horns did for a saddle. On this saddle we set a little figure or something that we could pretend was a man, or we tied something to hang down on either side like parfleche bags and played at moving camp.

Reed whistle (toys)

This "reed" is probably the softstem bulrush [*Schoenoplectus taberaemontani*, formerly *Scirpus validus*], which grows all over North America and thrives in water with some salinity (which would describe just about every slough in North Dakota). It can grow to 10 feet (about 3 meters) and has round stems.

Good Bird (vol. 9, 1910: 351)

When I was a boy we used to make whistles of reeds. Boys of eight to twelve years of age liked to do this. We loved to play with them. We cut a reed as in Figure 24. AA' is a slit cut in the reed. It would make a shrill noise. Such a whistle would last only for a day, for it had to be cut green. As it began to dry, we would wet it in our mouths to keep it green. After about a day it was too dry to blow.

In winter we made another kind of whistle of dry reeds as shown in Figure 25. A is a thin tongue of reed slipped into a slit. B and C are the end and cut openings.

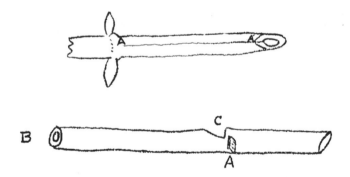

24. (*top*) A reed whistle. Drawing by Goodbird. (Courtesy of the Minnesota Historical Society)

25. (*bottom*) The other reed whistle. Drawing by Goodbird. (Courtesy of the Minnesota Historical Society)

FAMILY *Cyperaceae*–sedge family

GENUS *Schoenoplectus* (Rchb.) Palla–bulrush

SPECIES *Schoenoplectus tabernaemontani* (C.C. Gmel.) Palla–softstem bulrush

ELEVEN

Plants Used for Utilitarian Purposes

Cordgrass (thatching, lining storage pits)

HIDATSA NAME: *mika hatski* (*mika*, grass; *hatski*, tall or long)
LOCAL ENGLISH NAME: slough grass (prairie cordgrass)
BOTANICAL NAME: *Spartina pectinata* Bosc ex Link

Cordgrass is a coarse grass often found growing in thick stands where soils are wet most of the time. It can easily reach 5 to 6 feet (2 m) in height and has a sandpapery feel on the top of the leaf. Roofing and reroofing earthlodges required great quantities of this grass. Buffalobird-woman later refers to cordgrass as "blue grass."

Buffalobird-woman (vol. 20, 1916: 316)

This grass grows in the hills in wet places, thick; or at a steam's edge, or about ponds or lakes, or springs.

We used this kind of grass to put on the roofs of earthlodges and to line the walls of cache pits because it did not rot so quickly as other grasses. It lasted a long time.

This was the only use made of it. It was not used to make string for it is not, I think, very strong.

Buffalobird-woman (vol. 8, 1909: 59–60)

Grass for roofing the earthlodge

Over all [the sheathing of willows covering the frame of an earthlodge] was laid a thick matting of dry grass. A man stood within the lodge and, looking up, could tell if any places were laid thin as then light showed through. The grass was laid about six inches thick and covered all the lodge roof, both roof proper and also the lean-to [the *atuti*–the sloping sides of the earthlodge].

We cut the grass with a hoe. In old times, in the days of my parents, we used bone hoes. We always went out to find dead, thick grass that was long. This grass was to hold the final covering of dirt.

This much grass was used in an earthlodge of the ordinary size. We tied up two bales about two feet thick. These two bales were all that a woman could carry. It took one hundred bales to make enough to cover one lodge roof.

We brought the grass bales in on our backs or on the backs of ponies.

FAMILY *Poaceae*–grass family
GENUS *Spartina Schreb.*–cordgrass
SPECIES *Spartina pectinata* Bosc ex Link–prairie cordgrass

Buckbrush (brooms, medicine)

HIDATSA NAME: *macukaakca* (Goodbird does not know any
translatable meaning. *Macuka* means dog, but he thinks the word
is not a compound. GLW
LOCAL ENGLISH NAME: buckbrush (western snowberry)
BOTANICAL NAME: *Symphoricarpos occidentalis* Hook

Patches of buckbrush are scattered throughout the prairie, sometimes in dense thickets about thigh-high or as single plants but
usually in fairly close proximity to other buckbrush plants. They
are not a preferred forage plant and so tend to persist even in
areas which are being grazed. On archaeological sites in the Great
Plains, buckbrush will sometimes grow at the bottom of house
depressions where the soil is slightly more moist. At some sites
plants create a living map of features with, for example, buckbrush and pasture sage forming circles where houses once were
and other plants like cleome growing between the house depressions and forming a pink background with its flowers.

Buffalobird-woman (vol. 20, 1916: 243–48)

Buckbrush was used to make brooms. Every earthlodge had one
or two brooms used for sweeping the lodge floor and the ground
outside the door.

Brooms could be made at any time of the year, as needed. When
I made a broom in summer of green brush, I bound the bushes in
a bunch and hung them up, sometimes outside the entrance of the
door, till dry. Then I beat off the leaves and bound the bushes.
The larger and longer plants were chosen for making brooms.

These brooms were used in both the winter and summer earthlodges, but not in the tents when we were camping unless we stayed
there long enough to wear away the grass on the floor inside the tent.
Then a broom could be made at once from plants in the vicinity.

26. Buckbrush bundle. Drawing by Goodbird. (Courtesy of the American Museum of Natural History)

To make a broom I gathered good, long plants and laid them side by side to make sure they were of the same length. I then bound them together with a piece of thong so that the stems made a handle of about fifteen inches—enough for the two hands to grasp. We used no other kind of plant for brooms, but we did make them sometimes of buffalo hair.

In the earthlodge the place for the broom when not in use was near the left of the door, leaning against the wall of the lodge and placed handle downward. I do not know why.

A housewife made about two brooms in a winter and perhaps two, or possibly three, in the summer lodge the next season. A broom did not last very long.

The handle of the broom was bound about by the thongs in three places. We sometimes cut a good number of buckbrushes to make mattresses or beds when hunting—especially if the ground was wet or there was snow. This might be in the open air, or the beds might be piled around against the walls of the tent. Hunters did not always take a tent with them.

Such a bed was six or eight inches thick, and all the bushes were laid so that they were parallel to one another.

Buckbrush stems or sticks were used for arrows for boys. The boy's father would thrust the stem into the hot ashes of the fireplace; a green stem I mean. When it was hot, the father took a piece of skin in his hand and drew the hot stick through it holding it quite tightly. This peeled off the outer bark but left the inner yellow.

Such labor was always done by men—by the boy's father, brother, grandfather, etc. It was not a woman's part to do.

The arrows were unfeathered.

When a green stick was thrust into the hot ashes it went tsa-tsa-tsa with the heat.

The green inner bark of buckbrush was good for snow-blind eyes. A man would roll up a little ball of the inner bark of the buckbrush as big as the end of his little finger. Then he pulled hair out of a buffalo robe and wrapped it around the bark ball. This he dipped in water squeezing out the water lightly.

Then if there were two or three in camp who were snow blind, he went around and as they sat he squeezed a drop out of the bark ball into the corner of the eye of each patient. My father used to do this, and I know a Sioux woman who did the same.

The drop of liquid was rather strong to the eye.

The buffalo hair was put around the ball of bark to keep any little chip from falling into the eye.

When one was snow blind, the sunlight, or smoke, or wind hurt the eyes. Powder smoke hurt worst. A handkerchief was bound around to protect the eyes. Buckbrush medicine, used as above, was pretty good to help the eyes in snow blindness.

Many persons in the village used snow blind buckbrush medicine. It was in general use and was not at all a sacred medicine.

Another use of buckbrush was as follows:

All we Indians had pierced ears. When the ear was pierced with a sharp iron, a little tiny bit of the top of a buckbrush branch was taken, of a length equal to the thickness of the ear. This end was touched to a live coal to burn the ends smooth so that it would

27. A snare fence of buckbrush (and perhaps black sage). It is probably about 12–15 inches (30–38 cm) high. Drawing by Goodbird. (Courtesy of the American Museum of Natural History)

not lacerate the flesh as it was passed through the hole in the ear. The little stick was peeled and scraped smooth with a knife. It was left in the ear till the piercing healed.

This was universally done. No other wood or plant would serve the purpose. Why, I do not know.

Buckbrush was used for making the little fences in which snares were set to catch prairie chickens; black sage was also used, especially the dead and dry ones, for they were almost as strong as wood. Such snares were generally built in the spring when the birds are doing their spring dancing and the black sage can be found. The maker of the snare-fence used buckbrush if he could find enough, but filled in with black sage if needed.

Once we were out hunting and camped on Spring Creek (or Little Knife River) where there was some small boxelder timber. But we had cut all the dead and dry timber and had come to using buckbrush for fuel. It was March and the brush was live brush. Put on the fire, it soon burned out, but we found a way to remedy that. We gathered the brush into bundles five or six inches in diameter and tied the bundles with blue grass, [cord

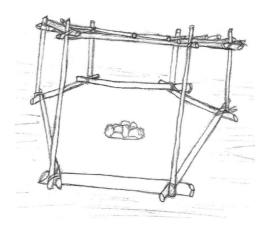

28. An improvised sweatlodge frame. At the center is the pit for the heated stones. Drawing by Goodbird. (Courtesy of the American Museum of Natural History)

grass, *S. pectinata*], which we wetted to make it pliable. These bundles we piled on each side of the door. This buckbrush fuel we used for a week, and no other kind, finding it very good. We even heated stones for our sweatbath with them.

We thought this plan of tying the buckbrush up into bundles ourselves. We never knew others to do it, for buffalo chips were plentiful on the prairie, and along the Missouri River there was plenty of wood.

The sweatlodge that we used here was unusual. The frame of the sweatlodge was made almost always of diamond willows [*Salix eriocephala*] because they have more spring; but if these were not to be had, some other wood was used, such as chokecherry tree or bear's berry (black haw) [nannyberry, *Viburnum lentago*].

Also at this place the ground was frozen, and as we could not find any long springy saplings we made a plan of our own. Six small logs were laid on the ground in a rough circle and uprights bound to them. A tent skin went over all.

In our party were Bad Dress and his wife Real Woman; Red Kettle and his wife Skunk Woman; Charging Enemy; Son Of A

Star, my husband; myself and my son Goodbird, about ten years of age. In another tent were Belly Up and his wife Has Many Young Ones and Belly Up's son Sitting Wolf, eight or nine years old.

Buckbrush berries were eaten. They were "ripe" in the spring, about March. Sometimes when we went out in the woods for fuel we women would gather each a handful of berries and eat them. They were not sweet, but we liked the taste.

The plants blossom after chokecherries blossom, late in June. Berries that follow are light colored, but the next spring when, as we say, they are ripe, they are turned black. [The berries are ripe in fall, but what Buffalobird-woman means is that by spring they have been through a few cycles of freezing and thawing which makes them more palatable.]

Prairie chickens eat buckbrush berries, green or ripe.

FAMILY *Caprifoliaceae*–honeysuckle family
GENUS *Symphoricarpos* Duham.–snowberry
SPECIES *Symphoricarpos occidentalis* Hook.–buckbrush, western snowberry

Cattails ("diapers," insulation for babies)

HIDATSA NAME: *hupatokike* (*hupati*, ear of corn—
like an ear of corn)
LOCAL ENGLISH NAME: cattail rush (broadleaved cattail)
BOTANICAL NAME: *Typha latifolia* L.

Almost every body of water seems to have cattails growing around it. For a plant which grows so well with its base submerged, it is quite capable of surviving the periodic droughts that are part of the Northern Plains climate. The broad-leaved cattails described here are losing ground (and water space) to the narrow-leaved cattail (*T. angustafolia*), which is an invasive and aggressive plant.

Buffalobird-woman (vol. 20, 1916: 322)

Cattail rush was used for packing around the bodies of small babes, but only in the winter, because its use was essentially for warmth. The down of the heads was used.

The cattails were gathered in the winter by the men. A man would break off the cattails from the stalks and thrust them into his robe above his belt under his left arm where he had made his robe into a big pocket.

The babe was covered with the down all about the hips and between the legs and down to the knees and bound up tightly in its cradle bundle. Small children frequently urinated, and for this reason cattail down was of real use as a protection. When a babe on a journey cried and made a fuss we knew that he was wet and cold and uncomfortable. We untied the baby from his bundle and took out the balled up down that was wet. When a babe's bowels moved, the cattail cotton formed a ball as when the babe wetted. The baby was then tied up again and stopped crying.

Now I will speak from my own experience. When the child needed attention, I untied its cradle bundle or wrappings in which it was bound the first year and packed the baby's body in the cat-tail cotton which I pulled off the cattail heads for the purpose. It took, I think, about ten heads for one application.

Now we used cattail down only when it was very cold and we were on a journey. We never used it in the earthlodge.

One application lasted two or three days. I would apply, say, ten cattails in the morning we started on the journey. At noon or eve the child perhaps would cry, and I knew it was uncom-fortable. I would open his bundle and remove the wetted part which would be made into a kind of ball—for the down would ball up with the wetting. But I would not put in any more down. But at the end of two or three days I had taken out so much of the down that I now had to replace it with some more.

We usually did not untie the baby's wrappings until we camped. Then, if I found too much of the down spoiled, I put some in that evening. But if not too much of it was spoiled, I let the unspoiled part of the packing remain as sufficient.

On such a journey, we carried along an extra supply in a sack, in the whole heads, and I would pull the down off the heads when needed.

Other packings for a babe

In the earthlodge we did not use cattail down, but buffalo-chips and sand. There were two kinds of buffalo-chips. One kind was flat and round. These we did not use. The other kind was dropped in little piles. These, dried, we used for babies in the winter time. The chips I powdered up with my hand. I laid down a cloth and put the buffalo-chip powder on it, then another cloth, and laid the baby on this. In old times instead of cloth we used skins, in the same way, however, both furred and un-furred (dehaired) skins. The buffalo-chip powder did not touch the baby's body.

I have heard that careless people used powdered rotten wood instead of chips; but the rotten wood we thought bred worms, and we never used it in my family. Sand was also used after being heated to keep the baby warm for the night. The sand was heated in the evening. Sand and buffalo-chips and cattail-down were not used in the summertime.

Skins when grown bad were thrown away, not washed. But if not bad, they were dried out and softened and used over again. When necessary a wet skin was taken out and a dry one put in its place.

When a child was two or three months old a skin was just bound around the hips reaching nearly from the navel to the knees.

We were careful of our babes in old times. Some families I have heard were careless, and their children broke their backbones in consequence. So it was our custom to put a piece of stiff raw-hide next to a baby's spine to stiffen it and protect it from strain.

FAMILY *Typhaceae*–cattail family
GENUS *Typha* L.–cattail
SPECIES *Typha latifolia* L.–broadleaf cattail

Boxelder (sweet and utilitarian)

HIDATSA NAME: *mitetadiki*
LOCAL ENGLISH NAME: boxelder or ash-leaved maple
BOTANICAL NAME: *Acer negundo* L.

Boxelders are often relatively short-lived trees that thrive in the earliest stages of plant succession on damp soils along rivers as well as in the woodlands found on the north slopes of hills especially in the badlands. They tend to grow rapidly but lose limbs easily and take on a very gnarled appearance before they are more than a few decades old. They are dioecious (male and female trees).

Although they do not look like maples at first glance because of their compound leaves, their winged seeds (samaras or, better yet, "owl knives") are distinctly maple. Boxelders were used in making basket splints and were a source of sweet sap in the spring.

Buffalobird-woman (vol. 20, 1916: 267–69)

This tree is one that increases very fast. Many grow in the timber hereabout and many more in the hills, for they grow fast. The tree is very quick growing.

There are two kinds, for just as with ash trees, one kind has a sort of seed blossom that we call "owl knives," and one kind has not.

In the spring, when the snow was melting but the leaves had not sprouted yet, we children would go out to a boxelder tree, break a branch, and tie a cup under the broken end. Sap would drop into the cup. Sometimes we cut a cavity in the tree and drank from it, cutting a tiny channel to which we put our mouths. This we did only when the south wind was blowing and then only at midday. The flow stopped in the evening.

We children did this. We drank the sap, and it smelled like boxelder wood. We called this drink *midaadaxi*. It tasted sweet. Both boys and girls would cut cavities in the trees to get sap.

29. Collecting boxelder sap. Drawing by Gilbert Wilson. (Courtesy of the Minnesota Historical Society)

Making baskets

Boxelder bark was also used for making baskets. The time for getting the bark was in the juneberry season. Then the bark peels off easily and is white and pure. In the spring when the leaves are small, the bark does not peel off the tree easily, some of it sticking to the trunk.

To get the bark I took an ax and cut a horizontal slit along the bottom of the tree near the ground. Then I peeled the bark in strips, peeling upward the length of the trunk to the branches. We never peeled downward.

As the bark was peeled off, the outside brown rind of bark fell off of itself.

The bark I took home and dried, hanging it on the drying stage, or sometimes laying it on the ground, but always with the inner side of the bark upward. That is, in drying the bark, the side that was exposed to the sun was just the opposite of the side exposed to the sun naturally on the tree. The side next to the tree is the best side and should be on the outside of the basket.

While drying, the bark must be kept from the rain for if wetted it spoils. After about five days, when it was dry, I took the bark down to the river and put it under the water in a long bundle, tied in two places with pieces of cloth and weighted down with a stone. When well soaked I brought the bark home and cut pieces into strips which I now rolled into rings about fourteen inches in diameter or into figure-of-eight bundles tied in the middle with a piece of bark.

These rings of bark I put away until I needed them for weaving baskets. The bark so prepared remained good for two years at least. I never tried them after keeping them longer than that and do not know whether they would be good or not. I made baskets at any time, winter or summer.

Boxelder wood makes a good fuel, but the sticks are apt to hurt the hand when one breaks them in his hands. Also the sticks are apt to spring about when one tries to split them.

Buffalobird-woman (vol. 12, 1912: 43)

The "root-post" carrier

Near one of the supporting posts around the edge of the lodge interior a post of boxelder was set into the ground with a root end up and the roots cut off. On these were hung the rawhide or braided ropes, sometimes several hanging on these roots. In old times, before we got nails from the whiteman, we hung saddles on the root carrier as it was set out far enough to allow saddles to be hung all around. This root-post was called *wideputai*. When

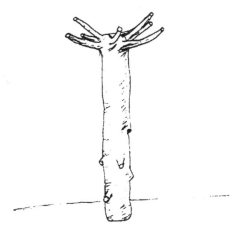

30. A boxelder root hanger. Drawing by Goodbird. (Courtesy of the Minnesota Historical Society)

anybody carries around a lot of things they call him *iwidiputsik-ets*—or say that he is "like that root-post set in the lodge." After nails came in we usually hung the saddles on a stringer [the horizontal beams supporting the lower ends of the rafters].

Baskets were usually hung on the root carrier. Bark baskets were always kept dry and so were hung up off the floor. Sometimes they were hung on the corral-posts, as the horses were not kept inside all the time, only on special occasions.

For more on uses of the boxelder, see also the sections on "Native Drinks of the Hidatsas" and "Wood."

FAMILY *Aceraceae*–maple family
GENUS *Acer* L.–maple
SPECIES *Acer negundo* L.–boxelder

Buffalograss (drying surface, embroidery)

HIDATSA NAME: *amauxihica* (*ama*, earth; *uxi*, antelope; *hi*, hair; *ca*, like. Earth that is like antelope's hair.)
LOCAL ENGLISH NAME: antelope-hair grass, as given by Goodbird, but I think it is the same as buffalograss in the neighborhood of the reservation. GLW [Buffalograss it is.]
BOTANICAL NAME: *Bouteloua dactyloides* (Nutt.) J. T. Columbus

Buffalograss does not tolerate being shaded and so does well in sunny, well drained soils where other grasses may have a hard time. It is a sod forming grass, and the area covered may vary from year to year as rainfall patterns change, but usually some of it manages to survive when the climate is hot and dry. Both the villages at the mouth of the Knife River and Like-a-Fishhook are just at the edge of the mixed-grass prairie, and one finds a mixture of short grasses and mixed grasses with occasional patches of tallgrass prairie grasses in places apt to accumulate a little more rainwater. Buffalograss has male plants, the flowers of which stick up above the leaves, and female plants, the flowers of which are enclosed in the base of the plant. Buffalograss also spreads by means of the vigorous growth of runners (stolons) when the opportunity arises.

Buffalobird-woman (vol. 20, 1916: 225)

When I was younger and it was the season to dry squash, I used to choose a patch of antelope-hair grass always on which to dry the solid slices of squash which could not be spitted on a stick. If we tried to put them on spits, the slices broke.

Also when dried meat got wet or was moist and was therefore likely to spoil, antelope-hair grass was good to dry the meat on.

Also when we dried squash blossoms we chose antelope-hair grass to dry them on.

Indeed this grass is good to dry almost anything on because the grass grows matted and thick and quite covers the ground almost like moss so that there is no earth exposed to soil the drying article.

We also used the roots of this grass for embroidery. These roots when fresh are not black but become so as soon as dried. The roots also are tough, and unlike those of some plants, they remain tough even after drying. When a woman made porcupine quill embroidery she put the black stripes in with these black antelope-hair grass roots. Other colors—red, yellow, and white—were made with porcupine quills, but the black was made with these roots.

FAMILY *Poaceae*–grass family
GENUS *Bouteloua* Lag.–grama
SPECIES *Bouteloua dactyloides* (Nutt.) J. T. Columbus–buffalograss

Big bluestem (toy arrows, hay)

HIDATSA NAME: *mika hicike* (grass red)
LOCAL ENGLISH NAME: red arrow-grass, red-grass (big bluestem, bluejoint, turkeyfoot)
BOTANICAL NAME: *Andropogon gerardii* Vitman

This is not exactly a common grass on what is now the reservation, but being a grass of the tallgrass prairie to the east, it does grow well where there is enough moisture or in patches widely scattered in the cover of the midgrass or mixed-grass prairie. Wolf Chief notes, "We used to go down in the flats where the red-grass grew and make grass-arrows." Although it is not always possible to discern the blue in the bluestem, it is not at all difficult to see this grass in the late summer and fall as "red-grass." Both big bluestem and little bluestem turn red or bronze, which sets them apart from other grasses.

Boys "playing at war" is essentially a universal in societies in which there is a chance that the males will actually participate in real warfare. In the film *Dead Birds* by Robert Gardner (1964), there is scene in which a group of Dugum Dani boys (highland Papua New Guinea) are doing exactly what Wolf Chief describes in his recollection of how he used red-grass (bluestem).

Wolf Chief (vol. 20, 1916: 307–10)

The stalks of this grass were used by boys of six to ten years of age for making arrows. These arrows were used in mock battles. For a bow we just used any green stick. A poor boy without a father had to use very poor bows; just anything he could get because he had no one to make them for him.

When we fought we mocked one another and cried, "Your bow is just like a basket standard" (meaning the bent willow branch that made the frame of a carrying basket). As we thus called we made the sound of contempt by vibrating the hand over the mouth or striking the fingers on the throat in a vibrating motion.

Sometimes the boy so mocked wept in shame. Especially if a boy was shot with grass-arrows and was hurt, and we saw he was almost weeping. We then mocked him and made him weep with anger.

I used to shoot grass-arrows with a bow of buffalo rib that my father made me.

Quivers for grass-arrows were made of gopher skins or prairie dog skins, or were handsomely made of deerskin.

A boy wore a quiver on his left side under his arm so that he could pull out the arrows with his right hand.

The bow that we used for hunting birds was longer and stronger than a grass-arrow bow. A grass-arrow bow, as I have said, was made of any green limb; but my bow for hunting birds, however, I would put away for the next time I might have a need of it.

We used to go down in the flats where the red-grass grew and make grass-arrows. These we divided amongst us so that we each had an equal number. Then we played at mock fighting.

Also in the fall we made shields of sunflower heads. Two holes were put through a sunflower head and a thong put through with sticks to hold it. This thong swung the shield on the left side, passing across both shoulders but not under the arm. Nor did it pass under the left arm like a quiver. This sunflower shield protected the left side.

Sometimes I would go to another lodge and call a boy to go with me to make grass-arrows and play at fighting.

Once I remember when I was nine or ten years old someone came and said, "Let us go and get grass-arrows." We went a little more than a mile and gathered up tall grass stems and made arrows and came back outside of the village to a place where we used to fight.

We had gathered up a number of boys, and we came close to one another and shot our arrows at each other, dodging the

31. A red-grass (big bluestem) arrow. Drawing by Goodbird. (Courtesy of the American Museum of Natural History)

arrows that we saw coming. Sometimes we charged against the other side. There would always be some boys that were smaller than others, and they could not run so fast. We older boys would catch up with the little fellows and shoot them with our arrows. It hurt where an arrow struck and often made the little fellows cry. The next day there would be a blue spot on the place where the arrow struck. That is the way we used to play.

This red-grass was also used to tie up squash seeds when women wanted to sprout them for planting.

Note: Specimens of red-grass were brought into the cabin at my request to show the size of the grass preferred for arrows. Wolf Chief says that the arrows were gathered the last of August and on into the autumn but that they were used fresh. They were never dried or kept. A typical specimen brought in had six joints and was five feet three inches above ground. To make the arrows, Wolf Chief broke the grass stem into sections exactly at the joints, nipping off the leaf that clung to each section so as to leave it like a feather to guide the arrow in its flight. "It won't go straight without the leaf," explained Wolf Chief. The specimen mentioned furnished five arrows. GLW

[For more on grass-arrows and boys using bows and arrows see the section in chapter 14 on arrows.]

FAMILY *Poaceae*–grass family
GENUS *Andropogon* L.–bluestem
SPECIES *Andropogon gerardii* Vitman–big bluestem

Common rush (mats, toys, dolls)

HIDATSA NAME: *matsatsuta*
LOCAL ENGLISH NAME: mat rush (common rush)
BOTANICAL NAME: *Juncus effusus* L.

Buffalobird-woman (vol. 9, 1910: 307–11)

About rush mats

Mat weaving was not a sacred thing. Anyone who wanted to learn had the right to learn and could go to a mat maker and learn, but not every woman in the tribe could make mats. It was not easy work, and not very many cared. If a woman wanted to make mats she could go and study the process herself and thus learn to make them. But it would have been hard for one to learn without a teacher.

I learned from my mother how to make them. In our village at Old Fort Berthold I think there were about five women and one man who made mats. The man's name was Red Stone, and he made great numbers of them. He had the mats in a circle all around in his lodge. He was the one who kept the Okipa objects at that time. He was a Hidatsa, but the Mandans did not find a man of their own tribe who pleased them so well as keeper as Red Stone.

In any lodge there would be one or two rush mats. Mats served in an earthlodge much as chairs do in a whiteman's house. All floors were of earth, and a rush mat was spread for one to sit on so that one's clothes would not get soiled. Anyone who went to visit in an earthlodge knew that he would be invited to sit on the mat. Often a mat was laid in front of the bed.

We learned to make rush mats from the Arikaras. They had learned it from the Pawnees. Our tribe had learned the art long before my lifetime. I do not know how long ago it was when the art of making mats was taught to our tribe. I began making mats when I was twenty years old.

The Hidatsas and Mandans no longer make these mats, and the art is almost forgotten. I do not think we have made mats for about forty years.

About forty years ago we began to use chairs and boards to sit upon. The Government put a saw in the village and sent men to use it. These men sawed square pieces about a foot high, and made benches. These seats were arranged around the four big posts in the lodge. In case a council was held, the seats were moved back to the exterior posts.

When a guest entered the lodge he always came in on the right side of the door. "Come in," we would say. "Sit in the *atuka*," the honorable place back of the fire. At this place there was always a mat ready. The owner of the lodge and his wife sat there a great deal.

Very often we ate on this mat. The husband and wife frequently ate from a bowl at the same time.

(Catlin pictures the Mandan women as never eating until after their husbands. Buffalobird-woman was asked if this was so; she laughed very heartily at the idea that the wife should wait on her husband like a servant. GLW)

The husband always had a butcher knife in his hand and cut off food for us women. We women would take up the pieces and eat them, of course.

Mats were never hung up on the wall; they were always laid on the floor.

Our moccasins had smooth soles and did not injure the mats when we walked over them. We left the mat on the floor all the time. We moved it a little when we swept.

The warp of the mat was always made of rawhide. We twisted grass into twine to string dry squash upon, but never to make warp for mats.

The Mandans and the Hidatsas both made mats. We could not distinguish the mats made by the Arikaras from those made by us. Our mats were just as good as theirs.

There was quite a difference in the goodness of the work because some mat makers did not do neat work, and their mats were loose and imperfect.

But in this region only the Arikaras and Mandans and Hidatsas made mats. Other tribes around here did not have enough home-life for such industries. They lived off the buffaloes and were always away camping and hunting.

Our mats were a little over three and one-half feet wide—the length of a rush of the most useable size. The length of our mats was about seven feet at the utmost or about the length of one of our beds.

Our earthlodges leaked a good deal. If in a rain a roof started to leak, the mats were rolled up and put away in a dry place and covered loosely with an old skin.

A mat was valued at one tanned robe, but a robe was often accepted that was not yet done.

I do not remember that the mat was ever put before the medicine shrine or sacred bundle that usually hung in the [back of the] earthlodge.

A half mile from Old Fort Berthold there is a lake. When we wanted to make a mat we went there and cut rushes and fetched them home, dried them, and began weaving. I often cut my own rushes. We mat makers did not keep rushes for winter work as we did bark to make baskets. A good mat should last several years at least.

Good mats should have the rushes cut in the fall and thoroughly dried. The time for making mats was in the latter half of September. A good mat, made of rushes cut in September when they were at their prime, should be golden. I have gone into the water after rushes at this time of the year when it was quite cold and chilly.

Buffalobird-woman (vol. 20, 1916: 321)

Making a rush doll

Rushes were also used for making dolls. Now a doll proper we call *maidakake*, but the doll we made of rushes was the "cradle-ornamented" or *itihakupe*. In the drawing by Goodbird, the original is shown after which the rush toy was modeled.

Also boys wove these rushes into whips or toy scourges. They played by striking one another with them without any kind of game.

In the Okipa ceremony one of the characters representing a hawk was armed with such a scourge but made of skin and called *maupakike*.

G. L. Wilson (vol. 20, 1916: frames 0087–0088)

[I am describing a doll made] by Buffalobird-woman assisted by Goodbird's wife and Hale's wife. The doll, to a whiteman's eye, looks like an Indian cradle, but Buffalobird-woman insisted it was not so regarded. It represents a babe, and she said, "We did not make cradles with hoods in old times although we know that other tribes did" (see Figure 32).

Buffalobird-woman says, "When a little girl of seven or eight years wanted a doll she asked her grandmother or some other friendly old woman to make it for her. It takes two to work, and often the little girl would hold the rushes while the older woman would do the weaving."

Buffalobird-woman (vol. 20, 1916: 148–49)

Rush scourges

MAUPAKIKE (A RUSH SCOURGE)

A stone hammer with which we pound bones we called *maupaki*. This "hammer" that I have made for you is called *maupakike* or "Like-a-stone-hammer."

32. The "ornamented cradle," which was the model for the reed doll. Drawing by Goodbird. (Courtesy of the American Museum of Natural History)

Boys sometimes chased one another with these. There was no regular game, but they did this just for fun. It was done about this time of the year (in August). Boys of about sixteen or seventeen years of age did this chasing of one another. Sometimes the boys made these hammers themselves, and sometimes they were made by the fathers and the mothers. The girls did not use them. Small boys just played with them, not hurting one another. Boys of sixteen and seventeen would chase and whack one another on the back.

Added by Goodbird:

On the last day of the Okipa, effort was made to imitate something. There were hawk imitators, antelope imitators, etc. All the men represented something. The hawks chased the antelopes with a kind of whip of leather with a great ball on the end.

These rush *maupakike* we considered imitations, and we did as the hawks did. We boys used to imitate everything whether it was sacred or not for we were reckless.

[The Okipa was perhaps the most important ceremony of the year for the Mandans and was a celebration of creation and thanksgiving.]

FAMILY *Juncaceae*–rush family
GENUS *Juncus* L.–rush
SPECIES *Juncus effusus* L.–common rush

Scouringrush horsetail (abrader, toy whistle)

HIDATSA NAME: *nokadaxiita iakoci* (ghost whistle)
LOCAL ENGLISH NAME: scouring rush, horsetail
BOTANICAL NAME: *Equisetum hyemale* L.

Horsetails typically grow in moist to wet soil and are often found around springs, seeps, and other damp places, although they can be found where one might least expect them, like in the Sand Hills of Nebraska. They can be very aggressive and will take over an area so that little besides horsetails can grow there. They are full of silica particles and hence "scouring rushes" because they can be used like sandpaper or a pot scrubber. They come apart easily at the joints, but then can be reassembled—great fun all. Wolf Chief's "whistle" would be more like an oboe reed—without the oboe. Wolf Chief's mixed testimony to the contrary, horsetails are poisonous.

Penn Veterinary Medicine at the University of Pennsylvania (2011) warns that for horses, "All members of this common genus should be considered poisonous." The toxic agent is an enzyme, thiaminase.

Wolf Chief (vol. 20, 1916: 295–96)

I do not know how to resolve the word *nokadaxi*. The word means "ghost," but *noka* is an old word meaning "people."

Ghost-whistles or scouring-rushes grow in the woods. They are good for fattening horses. Horses turned out onto the timber in wintertime sometimes come out fat because they find these rushes there to eat. Again, horses turned out into the timber get sick and, after two or three days, die. I guess from eating ghost-whistles, but I do not know why or how.

[Wolf Chief is just wrong in his notion that they are "good for fattening horses" but correct in his contrary observation that "horses turned out into the timber get sick and after two or three days die."]

Ghost-whistle rushes were used for smoothing or polishing

33. Ghost whistle (of scouringrush horsetail–*Equisetum hyemale*). Drawing by
Goodbird. (Courtesy of the American Museum of Natural History)

things. That elk-horn bow that I sold you was given its final polish with these rushes. When I was a boy I saw old men use these rushes for smoothing arrows. Both women and men used them for smoothing and polishing wooden bowls that they made.

I never saw a woman scour a dirty pot with ghost-whistle rushes. But I have heard that in the old days when they had clay pots, they saved a little of the liquid in which they had boiled corn, and this they poured in the pot and washed it to make the pot clean and more solid. They also cleaned bowls and spoons with a piece of hide with the fur on.

When I was a boy of about eight or nine years of age, I used to make whistles of these rushes. Now I have forgotten how I made them.

I used also to play with squash vine whistles. I would break off a hollow section of the vine, chew up one end somewhat, and blow into it.

Goodbird interrupting:

That was the way we made a whistle of ghost-whistle rush. I would break off a piece of the rush, chew up one end, and blow into it—putting the chewed end quite within my mouth.

FAMILY *Equisetaceae*–horsetail family
GENUS *Equisetum* L.–horsetail
SPECIES *Equisetum hyemale* L.–scouringrush, horsetail

Puffball (fire-making)

HIDATSA NAME: *micteda*
LOCAL ENGLISH NAME: puffball mushroom
BOTANICAL NAME: *Lycoperdon* spp.

From the information that Buffalobird-woman provides, I would guess that *Lycoperdon* spp. is the most likely puffball. Buffalobird-woman says slices were "about as big as the two joints of my first three fingers [about 2 inches wide] and "thick as a piece of blanket cloth," so it was probably a *Lycoperdon*. *Lycoperdon* have small pores in the top when ripe from which the spores escape. They typically grow in or along the edges of wooded areas.

Buffalobird-woman (vol. 20, 1916: 312–13)

The puffball was used in making a fire with flint and steel.

It was the man's job to make the fire. He cut a thin slice of puffball about as big as the two joints of my first three fingers and thick as a piece of blanket cloth. One side he rubbed well with gunpowder, wetted.

Every man had a flint case and also carried a steel made from an old file and shaped like a ring. This steel he used to make fire and for sharpening his knife as well.

I have often seen my father make fire with flint and steel. To make a fire he held three things in his hands: the piece of puffball, a piece of flint, and the steel shaped like a ring. He used a slice of the puffball the size I have described every time he made a fire. He did not carry many of these slices in his fire case.

The steel hung down, slung from a buckskin thong on his belt. The puffball was in a little bag or case with the flint.

To make a fire, my father took a little dried grass in his left hand, laid it on the little slab of puffball with the powder side up, and held the flint tight with his thumb. The sparks were stuck downward on the puffball slice, which caught fire with a swi-i-ish, and he folded the grass over the burning bit of puffball and shook it and waved it to the right and left as he held it in both hands till the grass caught fire.

He gathered the puffballs in the fall when they were ripe. I think he usually gathered three, when ripe, that is when the ball had opened. He cut the slices out with a knife. My father always cut them, not I. I do not know how many slices one puffball made.

This was the one use we made of puffballs. We did not eat them—they were not our food!

FAMILY *Agaricaceae*
GENUS *Lycoperdon*
SPECIES *Lycoperdon* spp.–puffball

Snakewood (arrows)

HIDATSA NAME: *mapukcaitawida* (snake his wood)
LOCAL ENGLISH NAME: snakewood (broom snakewood)
BOTANICAL NAME: *Gutierrezia sarothrae* (Pursh) Britt. & Rusby

Snakewood is found in the badlands immediately to the west of what is now the reservation. It is a plant that would seem dangerous to use for toy arrows because of its extreme toxicity, but arrows for boys was one of its principal uses. Lucero et al. (2006) analyzed the chemical composition of the plant, well established as having both medicinal and toxic properties, and extracted ninety-seven volatile compounds, most of which had never been shown to have been associated with the plant. Just what the function of this chemical complexity may be remains to be explained.

Buffalobird-woman (vol. 20, 1916: 335–36)

It grows among the rocks and on the white clay hills, especially in gullies made by the rains on the sides of badland hills. It is just a badland shrub.

This wood was used for making arrows for boys, of the sharp pointed kind. That is, the arrows were cut to a point and did not have a flint or iron head on them. The shrub has branches that are quite straight. The sticks were held over a fire after they had been peeled of bark, well oiled, and were straightened when heated. When dry the wood is a fine yellow. Bullberry [buffaloberry] wood when dry looks much like snakewood.

I have also heard that war arrows were made of this wood in old times. However, there is much about arrows that I do not know.

Snakewood we accounted a poisonous wood. If one was wounded by a snakewood arrow, the wound did not easily heal.

A boy named Umitsakic, or Well Painted Red (i.e., painted red all over his body—perhaps so named from having seen a red painted animal in a dream) was wounded in the cheek by another boy with one of these snakewood arrows. The arrow only scratched him, but the wound never healed until he was fourteen years old. He was six when he was shot.

We were all afraid of being wounded by this wood. I think it is called snakewood because it is poisonous like a snake.

Pipe cleaners were made from some hard wood like choke-cherry or sometimes bullberry. Snakewood was a pretty wood to use, and pipe cleaners might be made from it although I never saw this done.

(Wolf Chief has a pipe cleaner made from snakewood. Probably the name *snakewood* comes from its appearance. GLW)

Wolf Chief (vol. 22, 1918: 364)

About Snakewood

Spotted Horn had brought in some snakewood shafts. They were rather crooked.

"It is easy to straighten them for arrows after they are dry," said Spotted Horn. "Just warm them over the fire and bend them. But do not heat them too much!"

Wolf Chief then said, "We peel off the bark, dry them well, oil them, and heat them over the fire to straighten them. This must be done very carefully as snakewood shafts are easily broken."

I once adopted a Dakota Sioux named Red Leaf. Up the river in Montana in the Crow country he got drunk. He bought whiskey from a whiteman's store. It was at night, and while he was drunk he made his way back to his camp. Before he got there his friends heard him cry and talk. Some got up and went out and helped him to his tent. In the morning they found that he had been creeping through some snakewood and had run a piece of the wood in under the skin. They cut the skin and pulled out the piece of wood. But the wood had poisoned him, and in a short time his thigh and leg and belly swelled, and he died.

FAMILY *Asteraceae*–aster family
GENUS *Gutierrezia* Lag.–snakeweed
SPECIES *Gutierrezia sarothrae* (Pursh) Britt. & Rusby–broom snakeweed, snakewood

Goldenrod (a sign of the season, fishing gadgets)

HIDATSA NAME: *maiputsidi* (*ipu*, top + *tsidi*, yellow =
thing with yellow top)
LOCAL ENGLISH NAME: goldenrod
BOTANICAL NAME: *Solidago* spp.

There are many species of goldenrods, and they are not always easy to distinguish from one another. There are several species that are common to the midgrass prairie and, having little utility, they were probably all just "yellow tops." There are various insects that cause galls, but the most likely cause of the ball-shaped gall is the goldenrod gall fly, *Eurosta solidaginis*, which, in early summer, lays eggs on the stems of goldenrod and the developing larvae create more or less spherical galls on the plants' rapidly growing stems.

Buffalobird-woman (vol. 20, 1916: 301)

When the blossoms are all out in full bright yellow it is a sign that in our fields there is some green corn ready to eat. Such was a sign in old times.

[This might be true for *Solidago missouriensis*, which Stevens (1963: 274) says blooms in late July and is the earliest of the goldenrods to bloom. Other species are at their peaks in September. We picked our corn, grown from seed obtained from gardeners at Fort Berthold, North Dakota, at the end of August in both 2010 and 2011.]

Sometimes the stem of the goldenrod grows something like a light ball that we call *marotikadehe*. *Maroti* is the front part of a man's neck. *Kadehe* means rotten or sore. Translate—rotten neck.

This ball was used as a cork or bob to fish with. I never saw my father use this, but he was not much of a fisherman. Boys, I think, used it chiefly.

Goodbird:

We sometimes used these tough plants in the fall as sticks for playing *uakixeke*. I have often seen those balls used for fish bobs, chiefly by boys. The custom I hear comes from old times.

[The game of *uakixeke* is described in the "Wood" section.]

FAMILY *Asteraceae*–aster family
GENUS *Solidago* L.–goldenrod
SPECIES *Solidago canadensis* L.–Canada goldenrod

[I chose *Solidago canadensis* as a representative species. Probably all goldenrods were considered to be "yellow top."]

Prairie grasses as fodder

Wolf Chief specifies only what they didn't feed the horses, but the midgrass prairie has species like little bluestem, sideoats grama, hairy grama, western wheatgrass, and needle and thread that would be good winter forage. Fort Berthold, as noted, is just on the western edge of the midgrass (or mixed-grass) prairie, but along the Missouri River the midgrasses tend to prevail. There are scattered tallgrasses in areas in which a little more rainwater has accumulated—big bluestem (*Andropogon gerardii*) in particular.

Wolf Chief (vol. 22, 1918: 310–12)

Feeding horses dry grass

I have said that a drying-stage stood before every winter lodge and that this stage was used to dry meat on chiefly. But bundles of hay or dry grass were also kept here to feed the horses. I remember that on the stage before my father's lodge there were usually two or three big bundles lying on the stage floor in one corner, weighted down by a small log laid on them to stay them in the wind.

The women would go out with iron hoes and clean away the snow from places where the grass grew thick, put the grass up into bundles, and bring these in on the backs of pack horses. Men helped the women in this work. The grass sought was the prairie grass, not the red [usually referring to big bluestem] or river grasses [like prairie cordgrass–*Spartina pectinata*].

The women gathered this grass when convenient. They never kept a big store on hand. Just two or three bundles, as I have said, would be seen on the stage. It was always kept on the stage against need as horses did not readily eat hay that had been trampled or which dogs had fouled or run over. Piled on the stage floor, the grass was out of the dogs' reach.

The women went out for this grass at intervals. The grass was kept to feed only when the weather was stormy or at night, and at the end, say, of ten days, it would be exhausted, and three more bundles would be gathered. The hay was not fed to the horses regularly, only when there was a storm or enemies threatened. It was an emergency food.

We never put cottonwood bark on the stage floor as we put the bundles of grass, for the bark would freeze again. We fed no other kind of bark than cottonwood. It fattened the horses.

The women of our lodge used to go about a mile and a half [2.4 k] to the coulees for grass, returning with the bundles on their backs and the backs of horses. The bundles were tied up with rawhide ropes, of which we had always an abundance and when bound were about three feet long and nearly as thick. One such bundle was a load for a woman; four bundles were a load for a horse.

Sometimes also the women cut river grass with knives to feed to the horses. Later we learned to cut grass with a scythe.

When the bundles were brought in, a man or one of the women ascended the drying stage by the notched log ladder and a bundle was handed up to her. Sometimes they were piled on the floor, still unbound. Sometimes the bundles were unbound and a log laid over the hay for a weight.

If we wanted to feed the horses both hay and bark, the hay was laid on the ground under the rail first and the bark on top of it. If our store of hay was scant, a bit only was saved for morning feeding, none being fed the horses in the evening.

[For more information on the winter feeding of horses, see the section on the uses of cottonwood.]

TWELVE

Plants Used for Rituals or with Ritual Significance

The three kinds of sage

Sage (*Artemisia*) is a pervasive element of the Hidatsa world and of the Native American world throughout the entire range of these plants. Pasture sage from my garden has been a part of Dakota weddings, funerals, powwows, dedications, and sweatlodge ceremonies. Sage has its mundane uses as well—see the example of killing fleas in this chapter and the section on wild plums in chapter 3. It is a prominent element of the plains environment and, like the grasses, the cottonwoods, and the junipers, it evokes a vital essence of the Great Plains.

Wolf Chief describes three kinds of sage (actually four), and I have left this account as recorded by Wilson despite the confusion in Wolf Chief's descriptions, which cannot be resolved without the elusive missing specimens. Problems arise immediately when Wolf Chief describes four kinds of sage because one of them probably is not a sage (*Artemisia*) at all despite his identifying it as "lake sage." Other problems arise from the definition of a sage plant with flowers as being a different sage from one that does not have flowers. It is common to see both fringed sage (*A. frigida*) and pasture sage (*A. ludoviciana*) with and without flowers in any given year. Buffalobird-woman does say, "Except that it has no seeds ['no- top sage'], this is probably the same plant as the 'sage which has tops' (*A. ludoviciana*)." Sagebrush in the forms of *A. cana* Pursh and *A. tridentata* Nutt. is found in abundance in the badlands to the immediate west with *A. cana* also occurring along the Missouri River farther east.

Another problem for the reader (and the editor) is that people who depended so heavily on hunting did not need knowledge of botanical anatomy as precise as their knowledge of animal anatomy. At one point Wolf Chief described milkweed pods as being flowers (or Goodbird translated the Hidatsa term used by Wolf Chief as "flowers"). Wilson's botanical expertise was at best slight, and he simply could not ask the right questions to clarify which plants were being described.

HIDATSA NAME: *ixokataki*
LOCAL LNGLISH NAME: sage

Artemisia species described and discussed in this section:

BOTANICAL NAME: *Artemisia dracunculus* L.
LOCAL ENGLISH NAME: black sage (tarragon, green sagewort)

BOTANICAL NAME: *Artemisia campestris* L.
LOCAL ENGLISH NAME: Straight sage (common sagewort, field sagewort)

BOTANICAL NAME: *Artemisia ludoviciana* Nutt. (white sagebrush, prairie sage, also known as pasture sage)
BOTANICAL NAME: *Artemisia frigida* Willd. (fringed sagebrush)

The three [or four] species of sage are [as best I can determine]:

1) Hidatsa name: *Ixokataki aku ipudeca* (sage, kind of no top), and

Ixokataki aku ipumatu (sage, kind of has top),
Artemisia ludoviciana Nutt. (white sage or pasture sage)

Hidatsa name: *Miditia aku ixokataki* (Lake, kind of sage). Lake sage [unidentified but possibly a mint or lobelia]

2) Hidatsa name: *Ixokataki aku tsawutsi* (sage, kind of straight). This may be *A. campestris* L.

3) Hidatsa name: *Ixokataki aku cipica* (sage, kind of black)
A. dracunculus L. (black sage or green sage)

4) Wolf Chief does not describe *A. frigida*, known to the Hidatsas as "female sage" and otherwise as fringed sage

or prairie sagewort, which grows almost everywhere in North Dakota. In any year some plants bear flowers and others do not.

Wolf Chief (vol. 16, 1914: 299–303) (See also Buffalobird-woman's account from vol. 20, 1916: 344–45)

There are three kinds of sage on this reservation, as we Indians reckon them:

1. *Ixokataki aku ipudeca* (sage, kind of no top), or no top sage [*A. ludoviciana*]. This kind of sage was used as incense. It is the kind of sage that I (Wolf Chief) use when I go into the sweatlodge, and it is the kind of sage which Packs Wolf used when he let you fish with him one night in his fish trap. He used the sage to brush himself so as to ward off evil influences. Indian doctors often made a ball of this sage which they had in their hands when touching the sick or kneading the bellies of the sick. *Ixokataki aku ipumatu* (sage, kind of has top) is exactly like the kind just mentioned above, except that it has a head and bears seeds. [This is probably the same species as the preceding.] Each member of the Goose Society [a women's society] bore a bunch of this sage under her left arm. *Miditia aku ixokataki* (lake kind of sage), or lake sage, is again the same plant, but it grows in old water beds and damp places. Those in war parties often mixed it with white clay to paint their bodies to make them strong. After so painting himself, if one sweated, he smelled of the sage. This mixing of the sage with the clay was for a religious ceremony. Our belief was that it gave power to a man and made him a strong runner. My father used to burn this kind of sage as incense. He would set some to burning just behind his feet, and while he threw his robe over his head, he would inhale the smoke into his lungs. In a few

moments he would throw off the robe and vomit thick stuff from his stomach. He thought this kept him healthy and made him run strongly. I have tried it and found it is true; one can run a long way if he does this. [This does not seem to be a sage, so I will hazard a guess that it is *Lobelia spicata,* which, like other lobelias, contains a number of alkaloids and was used by other groups as a medicine plant for a variety of ailments (Moerman 2011). Its habitat preferences fit those given by Wolf Chief.]

2. *Ixokataki aku tsawutsi* (sage, kind of straight), or straight sage. [Possibly *A. campestris.*] This was the kind of sage which the Sun Singer wore when any sacrifice was made to the sun—one plant fastened on the back of his head. In the Big Bird's Ceremony, the singer would take a bunch of this kind of sage and tie it on his head. The singer of the Big Bird's Ceremony wore a crown of buffalo hair. In the spring we often found big tufts of hair shed by the buffalos. Sometimes in the spring when we were chasing buffalos, these mats or sheets of hair, not yet fallen from the animal's body, would flap and fly as the beast galloped along. The spots on the singer's crown (i.e., crown of buffalo hair) represent hail; they were made of wet clay. The wet clay was made into a little ball, and a string was put into it. When the ball dried, the string was held fast. By this string, then, it was possible to tie the little ball to the buffalo hair crown. The sage plume worn in the Kaduteta ceremony was just like it. In the Kaduteta ceremony, or ceremony of Old Woman Who Never Dies, the singer wore just one plant as a head ornament, and votive poles had a crown of the same sage.

3. *Ixokataki aku cipica* (sage, kind of black) or black sage [*A. dracunculus*]. This was the kind used by us boys when we wanted to burn our wrists. Black sage was used as a

ceremonial plant only in the sun offerings, so far as I recollect now. When a flag was offered, a little bunch of black sage was tied to the top of the flag pole. Black sage had an aromatic smell and was used a good deal as a perfume. The first named or no-top sage (*A. ludoviciana*) had a particular use in the eagle hunters' camp. If there happened to be in camp a menstruating woman, she was called in to the eagle-hunting lodge, and four balls of this kind of sage, painted red, were put on live coals of fire, and as they burned, the woman walked from one to the other and stood over them. The woman entered the lodge and passed to the right or east and stood over the burning sage with her blanket over her and inhaled the smoke. Then she passed in succession over the others. At each place she stood over the burning sage ball with her blanket covering her and inhaled the smoke. She then left the lodge. The directions north, south, east, and west [are defined ceremonially within the lodge]. Thus no matter how the lodge was oriented, the right hand, as one entered, was considered to be east. After the woman had thus purified herself, maybe the next day the hunters caught one eagle. If they did not have the woman inhale the black sage smoke, they thought all their sacred medicine objects would have their virtue weakened. I have told how we boys used to burn little bits of charcoal of the black sage to prove our courage. Our Indian doctors now use this same charcoal of the black sage to cure rheumatism. They put a bit of the charcoal, sometimes two pieces, on the rheumatic spot, on a joint, or the spine, the side, or wherever it may be, and, setting fire to the top, let the spark burn down to the flesh, exactly as we boys used to do in our play.

Pasture sage 1

HIDATSA NAME: *ixokataki aku ipuneca* (This is the *Artemisia ludoviciana* that has produced no flowers. *Ipuneca* means "has no tops.")
LOCAL ENGLISH NAME: pasture sage or white sagebrush
BOTANICAL NAME: *Artemisia ludoviciana* Nutt.

Buffalobird-woman (vol. 20, 1916: 302)

Except that it has no seeds, this is the probably the same plant as the "sage which has tops" (*A. ludoviciana*). The leaves and everything are the same.

Could be used instead of the "sage which has tops" to line the pit in which green plums were ripened.

Goodbird

In the fish-trap, Packs Wolf always brushed himself off after coming out from fishing with a bunch of this sage. Now the fish-trap was thought to be kind of alive and caught fish by its power. So when Packs Wolf went in, he came under the same influence—the magic that caught the fish. This power or influence he brushed off with the sage.

Sage is widely used in ceremonials, or was in the old times.

Pasture sage 1

HIDATSA NAME: *ixokataki aku hatski* (sage long) also *ixokataki aku ipumatu*, or sage that has tops
LOCAL ENGLISH NAME: pasture sage (white sagebrush)
BOTANICAL NAME: *Artemisia ludoviciana* Nutt.

Buffalobird-woman (vol. 20, 1916: 303)

If I wanted to make very plain that I meant this plant, I said *Ixokataki aku ipumatu*, or sage that has tops (or blossoms).

It is plentiful on the prairie.

This was the kind used as described to line little caches into which we put wild plums to ripen. Had not much other use.

Further comment by Goodbird

Remember, Wolf Chief told you that it was this kind of sage that Iron Eyes used in his sweatlodge.

[Both the "has top" and "no top" are probably the same plant, and it is widely used throughout the north-central states in many ceremonies despite Buffalobird-woman's disclaimer that it "had not much other use." One can find plants with blooms and without in any given year.]

FAMILY *Asteraceae*–aster family
GENUS *Artemisia* L.–sagebrush
SPECIES *Artemisia ludoviciana* Nutt.–white sagebrush

Pasture sage 2

HIDATSA NAME: *ixokataki aku ipudeca* (sage, kind of, no top) and
ixokataki aku ipumatu (sage, kind of, has top)
BOTANICAL NAME: *Artemisia ludoviciana* Nutt. (white sagebrush,
prairie sage; also known as pasture sage)

As noted, I am convinced that both are the same species. Pasture sage grows almost everywhere, more abundantly in the northern Great Plains and adjacent areas to the east. Besides the utilitarian uses given here, it has many ceremonial uses.

Buffalobird-woman (vol. 25, 1909: frames 58–60)

How hides were dressed for the tipi cover

My father's sister had the right of making tipi covers. She had bought this right from another woman. Afterwards I asked my (clan) aunt to sell me the right also. I tanned and dressed one good buffalo robe, very fine, and gave it to her for the right, and she taught me. She also taught other women the same craft but only those who paid her.

We made the tent covering usually of buffalo skins. These we knew how to tan or dress—soft like cloth. Cattle skins are hard to prepare thus.

A man wishing to have a tipi made brought in hides from his hunting and has them tanned by the women of his household. Skins that were to be used for a tipi cover should be soft and white, and I will now describe how we tanned or dressed them.

As I was taught by my (clan) aunt we tanned buffalo skins with buffalo brains and pieces of boiled flesh—the parts close to the small ribs near the kidneys. These were boiled with the brains and mashed together. When we had enough for the thirteen buffalo skins—a tipi of ordinary size—it took the brains of three or four buffaloes—we added one buffalo-horn spoon of bone-grease. This we put into a large earthen pot or into a vessel of wood. We now pulled grey sage [probably *A. ludoviciana*] and put it into the pot, mixing it all together.

A hide, dried and scraped, was laid on the ground upon an old tent cover. The hair had been removed with an elk-horn scraper bladed with iron. In old times we used flint blades. I have heard about them though I never saw them myself. In using the elk-horn scraper, the first two fingers were laid over the bend in the horn handle and down to the blade.

The hide was rubbed over its upper side with the sage and brain mixture, was turned over and its under side treated in the same way, and thus were all thirteen hides treated.

A tipi cover might have more than thirteen hides, but that number was usual.

About fleas

The animals identified here as kit foxes (*Vulpes macrotis*) were probably swift foxes (*Vulpes velox*). Kit foxes are no longer resident in North Dakota, although they may have been in the mid-nineteenth century, but kit foxes are generally found in the American Southwest. The swift fox is sometimes referred to as the prairie kit fox or the northern kit fox (National Museum of Natural History 2012), so confusion for all is understandable.

Wolf Chief (vol. 22, 1918: 247)

These fleas that are on kit foxes are just like those we now have in our houses but with this difference: when we trapped kit foxes to sell the skins to traders and packed the skins home on our packs, our necks and clothing fairly swarmed with fleas from the skins. Then when I got home I took coals of fire and upon them laid broad-leaved sage [*A. ludoviciana*]. This, of course, was in my tent. I squatted on my heels over the sage and drew my robe over my head and body, closing my eyes tight and holding my breath. I sat there for the length of time I was able to hold my breath. Short as this was, it was long enough to kill those fleas on me. I have done this often.

So too, when we trapped a kit fox we usually brought the carcass home and hung it on the pole over the earthlodge fire. The smoke soon killed the fleas and the carcass was then taken and skinned.

These fleas that we now have on our reservation, we have not had very long. They bite much harder than those that come off kit foxes. These latter did not bite hard, but one could feel them running all over one's arms.

The fleas we now have came to us about forty years ago, I think. People said at first, "Why, those are kit fox fleas," but they soon found out that they were quite different.

Only kit foxes were thus infested with fleas. Neither wolves nor coyotes had fleas, nor even the red foxes.

FAMILY *Asteraceae*–aster family
GENUS *Artemisia* L.–sagebrush
SPECIES *Artemisia ludoviciana* Nutt.–white sagebrush, pasture sage

Common sagewort

HIDATSA NAME: *ixokataki aku tsawutsi* (sage, kind of, straight)
LOCAL ENGLISH NAME: straight sage (common sagewort,
field sagewort)
BOTANICAL NAME: *Artemisia campestris* L. Unfortunately
this falls in the category of "probably" because of the sketchy
nature of the account.

Buffalobird-woman (vol. 20, 1916: 311)

Used to rub in the brains and meat with which hides were dressed.

Goodbird comments:

Used in many ceremonies to brush off the power or magic influences absorbed from the ceremony by the individual. A common custom in old days, but I have not more than a general idea of this—I cannot go into particulars.

Straight sage for sore eyes

Probably *A. campestris*. Describing a bout with snow-blindness Wolf Chief made the following observation.

Wolf Chief (vol. 18, 1915: 382)

My eyes hurt me for about three days. My father dropped some gunpowder in my eyes to cool them.

"This snow-blindness sometimes turns the eyeball white. If, after you are well, we find a white spot on your eyeball we will chew a piece of straight sage leaves mixed with charcoal and slip it in the eyes. This will take the white spot off." I have heard of this remedy from others of our tribe.

FAMILY *Asteraceae*–aster family
GENUS *Artemisia* L.–sagebrush
SPECIES *Artemisia campestris* L.–field sagewort

Black sage (medicine for arthritis, as "test" of boys' courage)

HIDATSA NAME: *ixokataki cipica* (sage black)
LOCAL ENGLISH NAME: black sage (tarragon, green sagewort)
BOTANICAL NAME: *Artemisia dracunculus* L.

Buffalobird-woman (vol. 20, 1916: 344–45)

I know of the custom of burning black sage for charcoal with which to burn the fingers. I have seen boys do this—burning black sage on their fingers for a test of courage. I remember Charging Enemy had the back of his hand badly burned this way. It was foolish and brought no good to any one.

Boys playing together in the evenings often made a fire and did this—the fire might be outside or inside the lodge. The boys' mothers used to try to stop it, but they wouldn't obey. Such boys were about ten or eleven years old.

Sick people who had rheumatism used to do this also. If one had rheumatism of the knee, the charcoal was burned on the knee over the place where it hurt.

Turtle Woman—who still lives—told me she once did this. She took a little piece about a quarter of an inch long and set it on end, upright, on the spot [that hurt] on the knee. To make it thus stand upright she wet the spot with saliva on the end of her finger. The charcoal bit was then fired and let burn down to the flesh. Turtle Woman said she did this several times at intervals, and after a few times the knee got well.

I once had a sore place on my spine at the back of my neck between my shoulders. I told my husband, and he got a bit of willow, just a little stick, and burned it to charcoal. I lay down, and he set the bit of charcoal upright on the sore place and burned it. It hurt me a great deal, but the soreness left me.

Wolf Chief (vol. 32, 1914: 229–30)

Black sage a remedy for rheumatism

I have told you how we boys used to burn little bits of black sage charcoal on our wrists to prove our courage. Our Indian doctors even now use this black sage charcoal to cure rheumatism. They put a bit of the charcoal or sometimes two pieces on the rheumatic spot—on the joint or the spine, or the side, or wherever the rheumatic spot was—and setting fire to the charcoal let the spark burn down to the flesh, exactly as we boys used to do in our play.

Wolf Chief (vol. 27, 1914: 39–46)

The fire test play

THE PLAY LODGE

The winter I was twelve years old my tribe made their winter village across the Missouri from Independence (Hill). To build winter lodges, the families for ten days had been cutting down trees and trimming the logs, leaving the trimmed off branches on the ground.

A number of us boys gathered one afternoon about two o'clock. I forget who first suggested it, but seeing the families at work building their lodges, we boys said to one another, "Let us build a lodge also."

We boys now built a play lodge. We gathered a quantity of forked branches and set them in a circle with tops meeting. We tied the ends of the poles above the door with young willow shoots. These made a pretty good substitute for cord. A boy would bite every part of a shoot to break the fibers to make it almost as pliable as twine. Our play lodge was uncovered, but one of the boys had a woolen blanket which we bound to the pole above the door with strips torn from the bottom of our cloth shirts.

Some girls had come around us, and we boys called to them, "Come over and let us play together," but they ran away. We boys chased, laughing, but they ran back to the camp.

There were, I think, eleven boys in our party. Many Sitting and I were of the same age, others were about nine, and some eight years of age.

We built a fire within, and as we sat playing, Many Sitting, a brother of Holding Eagle who was another boy of the party, said, "Let us burn our hands with charcoal." Sending one of the boys to a lodge for a live coal we then laid dry grass and twigs on it, and I was blowing the fire into a blaze when one of the boys struck my mouth in fun. When an Indian boy blows the fire, he does not pucker up his mouth as much as does a whiteman.

THE FIRE TEST

Many Sitting referred to charcoal of black sage, a plant we used in the fire test. We burned the stalk to a coal, which we smoothed with earth and ashes, the result being a little stick of charcoal. We broke the charcoal into bits about an inch long. By wetting one end in his mouth, a boy was able to make one of these bits of charcoal stand upright on his wrist. Then if a live coal was touched at the dry tip, the bit of charcoal took fire and burned steadily down to the flesh until all the charcoal was consumed. This fire test was a common thing for boys to suffer to show their courage.

GETTING THE BLACK SAGE

"Good," said the boys to Many Sitting's challenge, "but you begin the fire test yourself."

"That I will do," said Many Sittings. Some of the boys brought in some black sage plants, for it was now autumn and the stalks were dry. This sage made firmer charcoal than other weeds which burned quickly to ashes.

THE TEST

Many Sitting first put the bit of burning charcoal on his wrist. We boys knew that the flesh of the wrist is not very sensitive, and some cried, "Many Sitting, if you are going to be brave, you will burn your knuckle, not your wrist."

"Very well," said Many Sitting. "You may do so," and he held out his right hand.

The others took a bit of black sage charcoal on the mid-knuckle of the index finger and touched it with fire. It burned slowly down to the flesh. Many Sitting winced, but he did not jerk away his hand.

As the little spark burned closer and closer to his flesh, the other boys tried to frighten Many Sitting. "Look out, look out." they cried. 'It is going to burn you." I was one of those who so cried.

When the little coal of fire had burned out, Many Sitting said, "There, I think I am bravest of all of you. I do not think any of you have as much courage as I."

"I will excel you," I cried. "You have but one burn on your hand. I will beat you for I will let you put several burns on my hand."

"No, you can't beat me, you won't do it. You haven't the courage," said Many Sitting.

"Yes, I will," I said.

"How many burns do you want?" asked Many Sitting.

"As many as a jackrabbit's tracks," I answered.

"You can't do it, you don't dare!" cried the others.

"Yes, I do dare," I answered. "I am brave, you try me; I won't move."

"Yes, you will. You will wince," said the others.

"No, I won't," I said. "Here, you take the black sage charcoal and put bits on my left wrist just like the marks of a jackrabbit's tracks."

I had no reason to choose a jackrabbit's tracks for a pattern, but a jackrabbit as he runs makes many tracks, and this occurred to my mind. I thought I would taunt the boys by boasting that they might put as many bits of charcoal on my wrist as a jackrabbit makes tracks in the snow.

"Very well," cried all the boys, eager to try me at once.

They put fresh black sage in the fire, drew it out, and let it grow cold. They took seven little pieces of the charcoal and stood them on the back of my wrist. A little stick was thrust in the fire, and when the end caught in a bright glow, they drew it out and touched it to the top of each little bit of charcoal on my wrist, setting fire to them all.

"There," they cried, "look at him. He flinches already. He will wince. He wants to weep." They thus taunted me to break my courage. But I stood rigid and let the fire burn down to my flesh.

When the fire burned my skin, and the pain hurt me badly, I closed my eyes, but nothing more, and I stood without moving. Soon the fire was dead.

"There," I cried, "I show my courage. I am brave. Now see if any one else of you can stand more burns at one time."

Some of the boys tried the test, but none of them could stand the pain of several burns as I did. I remember one boy named Iron, younger than I, set two bits of charcoal on his wrist, but when the sparks reached his flesh, he knocked them off with his right hand, receiving but one burn.

OTHER PARTICIPANTS IN THE FIRE TEST

Of the other boys of our little company, I remember the names of the following:

IPATAKIXUPAC, SACRED WHITE FEATHER TAIL
AHUAWAIKIC, MANY SITTING
UWATAC, IRON
ITAHUXIEC, OLD MOUSE
WITEXADAXIC, LEAN BUFFALO
MAICUNEED, FLYING EAGLE

All these named and I think most if not all the others each put at least one piece of charcoal on his hand or wrist and set fire to it. But some as the fire burned close to the skin knocked off the burning charcoal. Three boys even wept with the pain. Sacred White Feather Tail, I remember, was one.

At the fire test I was sure to see what courage one had. Those who knocked off the burning coal or who yelled with the pain were chided by the others. Watching them I saw that some held the wrist steady but winced when the fire reached their flesh; others turned away their faces. I thought to myself that if I could but hold my countenance and not show evidence of pain, I should have this to boast afterwards.

Quite proud of my own accomplishment I tried to persuade some of the smaller boys to try the same (multiple) test. I chased some pretending I was going to make them submit to it. One or two of the smaller boys were frightened and ran home crying and play was broken up.

Evidently my companions told the other boys of the village how bravely I had suffered, and I was voted the bravest of all the lads of my age in the village.

WHAT WOLF CHIEF'S PARENTS THOUGHT

But in the evening I showed the wounds on my wrist to my father. My whole wrist was red.

"You foolish boy," said my father. "You gain nothing by doing that, and you may have a bad swelling on your wrist that may be hard to cure."

"Come here, and I will grease it for you," said my mother and she rubbed buffalo fat over the burnt wrist. What my father said and the redness on my wrist had now put me in a bad fright. And that night I was even more frightened. My wrist swelled up and pained me severely, and the glands under my arm swelled. I was kept awake all night by the pain.

FAMILY *Asteraceae*–aster family
GENUS *Artemisia* L.–sagebrush
SPECIES *Artemisia dracunculus* L.–tarragon

Fringed sage

This is probably *Artemisia frigida,* which is commonly called fringed sage. It typically forms a small mound of fine gray-green foliage. It is often found in association with pasture sage (*A. ludoviciana*). Most plants send up flowering stalks, but as was the case with Wolf Chief's description of *A. ludoviciana,* some plants may bloom and others not. Packs Wolf may consider the flowering stage to be a different plant.

Packs Wolf (from his account of the fish trap) (vol. 8, 1909: 46)

We Hidatsas call the short sage female or woman sage because it is short and has no seeds on top.

FAMILY *Asteraceae*–aster family
GENUS *Artemisia* L.–sagebrush
SPECIES *Artemisia frigida* Willd.–prairie sagewort, fringed sage

Juniper (Cedar) (bows, incense)

HIDATSA NAME: *midaxupa* (*mida*, wood; *xupa*, sacred)
LOCAL ENGLISH NAME: cedar tree (Rocky Mountain juniper)
BOTANICAL NAME: *Juniperus scopulorum* Sarg. *J. scopulorum*.
Thickets can be found on the more or less north-facing bluffs of
the badlands along the Little Missouri as well as individual trees
and clumps of trees scattered throughout the area (but generally on
northwest or north-facing slopes). These are very tough and enduring
trees, which made them desirable not only to the Hidatsas and
their predecessors but also to ranchers in search of fence posts.
Juniper wood is quite resistant to rot, which makes it valuable
to anyone using posts for whatever purpose.

Buffalobird-woman, Wolf Chief, and Goodbird
(vol. 20, 1916: 315)

These trees grow abundantly in the badlands along the Little
Missouri River on the sides of hills. But clumps of three or four
each will be found along the Missouri.

It is a medicine plant. The leaves are used, but it is not my
plant, and I have no right to doctor with it. So I cannot give you
information about it.

Goodbird:

It was much used for incense in old times.

Wolf Chief:

The wood was used for making bows, but these had to be backed
with sinew, for the wood is easily broken.

FAMILY *Cupressaceae*–cypress family
GENUS *Juniperus* L.–juniper
SPECIES *Juniperus scopulorum* Sarg.–Rocky Mountain juniper

Creeping juniper (incense, medicine)

Midaxupa aku okhatsedu (*mida*, wood; *xupa*, sacred; *aku*, kind of; *okhatsedu*, gliding—sacred wood of the kind that glides)

LOCAL ENGLISH NAME: ground cedar (creeping juniper)

BOTANICAL NAME: *Juniperus horizontalis* Moench

Creeping juniper is found near and on the tops of rocky bluffs in the badlands and in places where there is little to no competition from taller vegetation. It remains a plant with ritual uses, and like the common juniper it has blue berries (cones on the female plants).

Buffalobird-woman and Goodbird (vol. 20, 1916: 319)

Buffalobird-woman:

A medicine plant, but it is not mine to use, so I cannot explain its use.

Goodbird:

I have heard that this plant was used to mix with other plants to make medicine, but I do not know what they were. It was the leaves that were used.

Like the tall cedar it was used for incense. Incense or smoke made before sacred objects was much used in old times. It is a sacred plant.

FAMILY *Cupressaceae*–cypress family

GENUS *Juniperus* L.–juniper

SPECIES *Juniperus horizontalis* Moench–creeping juniper

Prairie sandreed (ceremonial ornament)

HIDATSA NAME: *mikatsatsa* (*mika*, grass; *tsatsa*, an old word meaning, Buffalobird-woman thinks, growing in a bunch)
LOCAL ENGLISH NAME: (none given) (prairie sandreed)
BOTANICAL NAME: *Calamovilfa longifolia* (Hook.) Scribn.

There are localized areas of sandy prairie just north of Crow Flies High Overlook on the Missouri River near New Town, North Dakota, where there are stabilized sand dunes and an occasional blowout providing a good habitat for prairie sandreed. It can also be found growing along the edges of high stream banks where the soil is sandy and too dry for many other grasses. This is a tall and easily noticed grass.

Wolf Chief (vol. 20, 1916: 297–99)

This is the kind of grass used in old times for head ornaments. Adapozis had such grass in his hair, as you will remember, in the story of Burnt Arrow. [Adapozis was one of the founders of the Hidatsa people, and a brief account of this can be found in the chapter on arrows.]

A man named Red Stone also used to wear a little bunch of this grass just over his forehead, thrust forward because he had dreamed of this in a vision. When he went to war he always put grass in his hair thus. Also I have heard that in old times men who went on war parties put this grass in their hair.

On a war party, he of the spies who first spied the enemy received the right to wear a head ornament of this grass. So also of the second, third, and fourth, but not the fifth, for only the first four counted.

If any other young man in camp spied out the enemy, he had the right to the same honor mark, but this was seldom, of course, because only those appointed as spies usually had a chance to spy

34. A sandreed (*Calamovilfa*) hair ornament. Drawing by Goodbird. (Courtesy of the American Museum of Natural History)

the enemy. The basis of the right to wear the ornament was that the wearer was a member of a war party under a regularly recognized leader. If a young man, for example, had gone out from the village and was herding horses, saw enemies, and returned to the village and reported, he did not receive this honor mark.

When the war party fought, and one of the four (who had spied on enemies and so wore the grass ornament) struck an enemy, he had the right to paint a red band across the middle of the grass ornament.

If a young man had spied the enemy once, he wore one grass. The grass was simply thrust into the hair by the stem if the wearer expected to use it only temporarily, or it could be bound to a little stick and thrust down and could be taken out at night.

For every time that the man spied out the enemy, he added another grass to the little stick forming a fan shape. I, Wolf Chief, for example, have spied out the enemy five times. A little bit of hair was tied tightly with a string or thong and into this the sharp stick was thrust. In the drawing, the grass appears more tassel-like than is usual in the real specimen.

35. This is probably Wolf Chief with his hair ornament. Drawing by Goodbird. (Courtesy of the American Museum of Natural History)

But I do not have the right to paint a red band on my grass orna-ment, for I never struck an enemy on the same war party on which I spied the enemy. For the right to paint the red stripe, one must count coup on one of the same enemies he had spied out and in the fight that immediately followed. It must always be the same enemies, the same day.

One put on the red band simply by wetting the paint and touch-ing the grass midway and rubbing slightly between a thumb and forefinger wetted with red paint.

One painted a band across each grass which represented an enemy spied out and an enemy struck in battle immediately fol-lowing. When one had won four grasses by spying and also four strikes on enemies in battles respectively following, he had earned four striped grasses red all over. Only a very few great warriors ever had so many such honor marks as to have earned the right to paint four grasses all red.

This was, however, only after four spy-and-strike grasses had been won. If a man spied and so won one grass and then won a strike also, if he now painted the grass red all over, people would have made a fool of him; they would have called him a dog, or crazy.

I have never done this, but I have heard that other people when they found a porcupine would break off a few tops of the head ornament grass and dart them with the hand like a spear at the porcupine. Then they killed the porcupine and found the quills good and straight and not bent or broken.

FAMILY *Poaceae*–grass family
GENUS *Calamovilfa* (Gray) Hack. ex Scribn. & Southworth–sandreed
SPECIES *Calamovilfa longifolia* (Hook.) Scribn.–prairie sand-reed

Bittersweet (sacred associations)

HIDATSA NAME: *maxupa miaita matsu* (*maxupa*–sacred, god, equivalent to Sioux wakan; *mia*–woman; *ita*–her, his, its; *matsu*–berry)
LOCAL ENGLISH NAME: Sacred Woman's berry (American bittersweet)
BOTANICAL NAME: *Celastrus scandens* L.

Although this plant usually grows as a climbing vine, it will also sprawl. The female plant is very striking in the fall as the leaves turn a bright yellow and the orange outer part of the fruit splits open to display the red berry within.

Wolf Chief (vol. 20, 1916: 342)

The Sacred Woman's berry plant has red berries when ripe, but they turn orange in winter. They grow on the vine, which climbs a tree in the woods. In fall the leaves turn brown and yellow. We never ate the berries. Perhaps they are good to eat, but we Indian people were afraid, for we thought them sacred. As we understand it, these were "the Sacred Woman's berries from old times." I do not know how many leaves were in a cluster, for we were afraid of the plant and let it alone.

There were two sacred women, one up in the sky, one on earth.

FAMILY *Celastraceae*–bittersweet family
GENUS *Celastrus* L.–bittersweet
SPECIES *Celastrus scandens* L.–American bittersweet

THIRTEEN

Sources
of Wood

Wood as a resource

The Hidatsas, living as they did in the Northern Great Plains, were faced with the fact that many resources used by their ancestors from the east and more or less taken for granted were either different, scarce, or absent. The most valuable plant resource, and one that was consumed in vast quantities, was wood. The Great Plains, both North and South, were, by definition, grasslands–prairie.

The eastern periphery was tallgrass prairie with big bluestem and Indian grass being prominent but by no means the only tallgrass species found. These grasses would often grow to a height of 6 feet (2 m) or more. As annual precipitation diminishes to the west, the tallgrasses were largely replaced by shorter grasses such as the wheat grasses, little bluestem, grama grasses, and the needlegrasses, as well as many other species that were more drought tolerant. This is the mixed-grass prairie. Still further west is the short-grass prairie, with buffalograss being perhaps the most familiar of the many species found there.

There were essentially no trees with the exception of those growing on the floodplains of the larger rivers of which the Missouri River was by far the largest. Villages were situated along rivers because floodplains had fertile soil; a soil that could be tilled for gardens with digging-sticks and bison scapula hoes. Rivers also were a reliable source of water in country that is dry and has many springs that are intensely alkaline. And, very important, rivers were bordered by trees. It cannot be stressed just how important wood was to the villagers (Fawcett 1988 and Griffin 1977). No village could be built without a significant source of construction-sized timber for houses and palisades. The necessities of daily life, like cooking and keeping warm in cold weather, also required substantial amounts of wood. The collection of firewood required a great deal of time, particularly, but not exclusively, by the women.

The distribution of trees was determined by a combination of diminished precipitation and recurrent prairie fires. This limited trees largely to the floodplains along rivers and north-facing slopes, where the heat and drying action of the sun was less intense and soil moisture more abundant. The most common trees originally found on the floodplains were the cottonwoods; less frequently found were ash, American elm, peach-leaved willow, and boxelder. Various other willow species, some far more useful than others, grow primarily on river banks and sandbars.

On north-facing slopes there is a mix of small trees: juniper, boxelder, chokecherries, juneberries, and ash. Raspberries, currants, and poison ivy, as well as numerous other plants, were part of the understory—a source of many of the plants used by the Hidatsa.

As he was excavating the butt end of an oak post from the center of a lodge near Stanton, North Dakota, just north of Bismarck, archaeologist Donald Lehmer explained the sequence: first to be used up were the bur oaks, because they were both durable and strong; then the junipers, because they were durable and adequate; and finally the cottonwoods, because they were all that was left (personal communication 1971a). Cottonwood is neither durable nor is it a particularly good firewood because of its low density. But it was the most readily available, and in fairly dense groves along rivers it tends to grow tall and straight.

The annual flooding of the Missouri and its tributaries (like the Knife, Yellowstone, and Little Missouri Rivers) once swept large quantities of dead cottonwood trees and branches downstream, some of it onto the banks and the sandbars that emerged as river levels dropped. This driftwood was highly prized for its superior resistance to rotting and because it was a lot easier to have wood delivered to the village than it was to transport it from ever-increasing distances as local sources were used up.

Construction of houses and fortifications

Hidatsa lodges, although called earthlodges, were constructed almost entirely of wood covered by a roughly 4-inch (10 cm) layer of well-tamped earth (see Wilson 1934, and the appendix in this book by his brother, Frederick Wilson, for exceedingly detailed accounts of earthlodge construction). Each lodge required as many as two hundred trees, ranging in size from 12 to 14 inches (30–35 cm) in diameter and 14 to 16 feet (4.0–5.0 m) in length for the four center posts and the four stringers they supported, plus twenty-four to thirty posts and stringers 8 to 10 feet (2.4–3.0 m) in length around the periphery of the lodge and 5 to 7 inches in diameter by approximately 20 feet (12–18 cm by 6.0 m) in length for the hundred or so rafters that supported the roof (Wilson 1934). A lodge lasted about seven to ten years, at which point rebuilding was necessary. Although much wood was recycled, a lot of it was unusable for anything other than firewood (Wilson 1934 and Scullin 2005: 32–50). Typical villages might have between twenty and eighty such lodges. Forty lodges would require about eight thousand trees every seven to ten years. And although some of the wood could be recycled, most could not—except as firewood.

Furthermore, many of the villages built along the Missouri River in what is now North Dakota and South Dakota were protected by palisades of varying length and elaboration. Some, like Huff, Black Partisan, and Double Ditch, had bastions constructed at regular intervals for even greater security. This type of construction requires an extremely large number of logs (roughly one 5- to 10-inch diameter log per foot [12–25 cm diameter per 30 cm length]). Just as the timbers of a lodge would be disintegrating after seven years, so also would the timbers used in the palisade.

At Double Ditch, a Mandan village just north of Bismarck, at least four palisades were built around the village at different times. The cutting and salvaging of timber never stopped. The outermost fortifications were the most elaborate, with bastions

36. Wood framing of an earthlodge. Drawing by Frederick N. Wilson. (Courtesy of the American Museum of Natural History)

constructed at regular intervals. Subsequent smaller fortifications were built to a simpler plan that consumed significantly less timber (Ahler and Geib 2007: 442–51 and Swenson 2007: 254–55).

Firewood

Because Hidatsa diets consisted largely of bison and corn, virtually all food was cooked, and firewood was in constant demand in every household. During the winter, villages broke up into smaller units and the people constructed seasonal settlements in the wooded areas along the Missouri River, where they sought shelter from the winter weather, a source of water, and readily accessible timber. Lodges in these settlements were considerably smaller than summer lodges. Winter lodges were about 25 feet (7.5 m) in diameter as opposed to 45 feet (14 m) for the summer lodges, but each still required many trees. Firewood demands would have been great because winter temperatures often dropped to zero degrees (-18 C) and considerably below. Earthlodges were far from snug. Every lodge had to have a large smoke hole in the center of the roof. This also admitted the only light other than that cast by the fire.

The Hidatsa villages at the mouth of the Knife River were occupied for generations, and even the last earthlodge village at Like-a-Fishhook was occupied for four decades (1845 to 1885). The constant necessity of cutting and gathering wood was an immense stress on both the wood resources along the Missouri and the people who had to do the cutting. The acquisition of iron tools early in the nineteenth century and the arrival of wood-burning steamboats shortly thereafter further intensified the harvesting of wood. Cutting wood with iron axes rather than with stone axes was far easier but also more dangerous as firearms became more common among all the various groups living in the Northern Plains. Woodcutters had to go farther and farther afield, which made them more susceptible to attack.

This was the period during which Buffalobird-woman and Wolf Chief lived: the iron age. Their knowledge of stone and bone tools was, for the most part, what they had heard from the older members of the group.

The scarcity of wood

Wolf Chief was about to leave the village for a hunt and is describing the care and sheltering of the horses. During times of trouble and bad weather, the best horse or horses were kept inside the earthlodge in a stall on the right as one entered; in times of hostilities a corral was assembled on the left side of the lodge adjacent to the entryway so that the other horses were more secure from theft. Even this simple construction was complicated by the scarcity of wood in the vicinity of the village.

Wolf Chief (vol. 14, 1913: 443)

In our family we used to keep the posts of the war corral, four or five of them, standing fork upward on the right-hand side of the lodge on the outside. The rails for the war corral were laid lengthwise on the ground beside these forked posts. I do not mean that this was done in peace times, but we looked upon the corral only as an emergency corral, to be put up in times of extreme danger and taken down again when the danger was over. The floor was then cleaned and swept. We had preserved these rails and posts because they were hard to get. We had to go two miles above the river and cut them, returning in a boat and floating the rails in the current.

Cottonwood (architecture, food for horses, sweet)

HIDATSA NAME: *ma'ku*
LOCAL ENGLISH NAME: cottonwood (plains cottonwood)
BOTANICAL NAME: *Populus deltoides* Bartr. ex Marsh. ssp. *monilifera*
(Ait,) Eckenwalder

Cottonwood trees are the most abundant species in the gallery forests, along the larger rivers, and particularly along the Missouri and Little Missouri. Both have broad floodplains across which the river continually scoured new channels and formed new land as it consumed older land (before the dams were built on the Missouri in the 1950s) . Cottonwood trees, by virtue of their production of millions of fluffy seeds that are easily transported by the almost constant winds of the plains, were among the first seeds to fall on the mud or moist sand of new and fertile areas. The Missouri River valley was lined with groves of cottonwood trees, each grove of a fairly even age and of a density such that most of the trees grew both quickly and straight, which was ideal for construction.

In the grasslands of the Great Plains, having a source of good timber was extremely important, and between the building and rebuilding of houses and the construction of protective palisades around many of the villages, these trees were consumed by the hundreds of thousands. There is also the disadvantage that cottonwood rots rather quickly, so that demand was constant and supplies were limited. Furthermore, cottonwood trees are quite "soft" and more easily worked with fire and stone tools. In the nineteenth century the use of both the branches and the inner bark of cottonwood to feed horses during the winter further contributed to scarcity in some areas. (For further information see the section on "Uses of wood.")

Wolf Chief (vol. 20, 1916: 249–50)

Cottonwood trees are very abundant in the woods along the Missouri and much less abundant in the hills. But they are abundant along the Little Missouri for 150 miles upstream.

We cut down cottonwood trees and fed the horses on the top branches in winter time. They got fat on it. [Doubtful.]

Cottonwood was a useful wood. It was used in building earth-lodges and corn stages and corrals for horses such as we made inside the lodge. Also it was commonly used to make the fort or palisades that surrounded the villages in old days, as around our village at Like-a-Fishhook bend.

In early summertime, about July, the bark was sweet, and sometimes one would take a knife and peel back the bark and chew the inner bark and scrapings from the wood immediately under the bark. These scrapings and the inner bark were quite sweet. Young and old alike did this. It was quite good to the taste!

We made tent poles of young cottonwood trees. They were cut in June when they were easily peeled of bark. And, being light, they were good for the two runners of a dog travois. Also cottonwood was a fine wood to burn for fuel. [Wolf Chief must be thinking about accessibility because cottonwood, being a low-density wood, is not good fuel (Slusher 1985: 2).]

We gathered up the dead wood and thick heavy pieces of cottonwood bark that floated down the Missouri for fuel to heat stones for a sweatbath.

Also at night a woman took a piece of bark that was heavy and thick and that for this reason she had saved and buried it in the ashes after setting it afire. It smoldered all night, and in the morning the fire was started with it. Our way then was much better than the coal we now use.

Cottonwood bark for horses

Wolf Chief was recalling a winter camp in which the family stayed when he was about eight years old (about 1857). A major task was to tend the horses, who had to be fed most of the winter. The acquisition of horses by the Hidatsa occurred at about the same time that they had abandoned stone tools for iron, the late 1700s and early 1800s.

Wolf Chief (vol. 22, 1918: 306–9)

I do not know how many lodges were in our winter camp. I remember we camped in three different places in the timber, and I would guess that there were perhaps ten to twenty lodges in each of the three camping places. There were then—and again I am guessing only—perhaps about two hundred horses in the tribe, all or very nearly all of them ponies. I think every lodge owned some ponies, but the number varied in the families. Not all the best horses were kept in the family's lodge at night—certainly no more than ten at the most. These ponies were confined by a railing in a pen or corral inside the lodge. The less valuable part of the village herd was left out in the hills, and as I recollect there were more ponies left in the hills than were driven into the lodges at night.

We fed the lodge-kept ponies cottonwood bark and the tops and small branches of the same wood.

Usually the women of the lodge went out in the afternoon to cut cottonwoods and would cut down two or three trees. Trees, say a foot thick (30 cm), were cut down. The rough outer bark was cut off and the green inner bark was stripped off. These strips, many of them as long as my arm and as broad as my hand, were fetched to the lodge and piled near the fire to thaw. The women also fetched in the smaller branches lopped from the tree. The bark and twigs and small branches they would fetch to the lodge and make them ready to feed the ponies at night. As this was the fourth year we had wintered at this site which we called Buckbrush-pit Timber, the cottonwoods were becoming pretty well cut off about our camping place.

The women were careful to clean up the cut-down tree of both bark and small branches, because if they did not, any horses that were in the woods were sure to go to that tree and browse on the bark and branches and the woman's labor would be lost.

In the evening, before dark, the bark and branches or twigs which had been thawed out by the fire were piled under the rail of the corral for the horses to eat. This was about sunset. The ponies ate at night, but part of the bark and cottonwood branches were saved to feed in the morning.

The horses stood facing the fire, and the cottonwood bark and branches were piled under the corral rail for the ponies to eat. The twigs and smaller branches they ate, and branches as thick as my wrist they stripped of bark with their teeth. The twigs were cut off and piled by themselves. This was done so that the thicker branches, thus lopped of twigs and smaller branches, could be more easily stripped of their bark by the ponies' teeth.

This custom of feeding cottonwood bark in the lodge was true only of our more valuable horses. The year I am telling this of, we had nine ponies in my father's lodge. But this number might easily be broken. A sister or an aunt—either a blood aunt or a clan aunt—might present valuables to the owner and receive a pony as a gift in return. All the nine ponies which made up my father's herd he brought into his lodge to feed and tend because my father was a provident man and was wise enough to bring his horses into the lodge every night.

Wolf Chief and several other men were on a hunting party and had just broken camp.

Wolf Chief (vol. 22, 1918: 204)

When we had put on our clothes and had just entered the timber, we were all amazed to see an immense tree. We had never seen such a large tree, and we all stopped to look at it. In a fork of the tree ten or fifteen feet from the ground there were some kinnikinnick bushes [red-osier dogwood, *Cornus sericea*] growing. "What a big tree," we said. "Let us measure it." Five of us stood with our arms outstretched and fingers touching, and we could not reach around the tree. The sixth man, standing in the circle reached the fingers of the first with his second shoulder. It was a cottonwood tree and the biggest I ever saw.

[See chapters 13 and 16 for the uses of cottonwood, and for more on feeding horses dry prairie grass in the winter, see chapter 11.]

FAMILY *Salicaceae*–willow family
GENUS *Populus* L.–cottonwood
SPECIES *Populus deltoides* Bartr. ex Marsh.–eastern cottonwood
SUBSPECIES *Populus deltoides* Bartr. ex Marsh. ssp. *monilifera* (Ait.) Eckenwalder–plains cottonwood

Ash (tools)

HIDATSA NAME: *micpa*
LOCAL ENGLISH NAME: ash tree (green ash)
BOTANICAL NAME: *Fraxinus pennsylvanica* Marsh.

As Buffalobird-woman notes, ashes are commonly found in west-central North Dakota, but they seldom grow very large. Her distinction between trees that bear seeds and those that do not is the same distinction made by botanists. Ashes are dioecious, having male and female flowers on different trees. Along with junipers and boxelders, they are often found on the cooler north-facing slopes.

Buffalobird-woman (vol. 20, 1916: 264)

Ash is a rather common wood along the Missouri. There are two kinds of ash, and they are both exactly the same tree except that one has blossoms and one has not. Both kinds are found in the same part of the woods. The greater number are without flowers, and only a few bear blossoms. I think that those that are barren remain barren every year.

We did not use ash wood for a great many purposes. It is a hard wood, hard to cut, but it makes a hot fire when used for fuel.

Wedges were made of ash. We used them to split planks of cottonwood. These wedges were cut from both green and seasoned wood.

Although I never saw it myself, we hear from old times that wedges were made of buffalo horns. To make it solid, a piece of ash was driven into the hollow of the horn.

The hoop of a travois [the platform on which things were carried], whether for dog or horse, was made of ash.

Corn mortars and the pestle or pounder were made of ash also.

[One of the most important uses of ash wood was for the manufacture of gardening tools, particularly digging-sticks. Digging-sticks were multipurpose tools and were made in various sizes according to their intended use. See chapter 14 "Uses of wood" for more information on digging-sticks.]

FAMILY *Oleaceae*–olive family

GENUS *Fraxinus* L.–ash

SPECIES *Fraxinus pennsylvanica* Marsh.–green ash

Peachleaf willow (utilitarian)

HIDATSA NAME: *maxoxica*
LOCAL ENGLISH NAME: Indian willow (peachleaf willow)
BOTANICAL NAME: *Salix amygdaloides* Andersson

Peachleaf willows can grow to a height of 50–60 feet and, as noted by Buffalobird-woman, are usually found along the banks of rivers and streams. They are the only willow trees. They were used mostly as a source of bark splints, as frames for baskets, and for bullboat frames. The following is an excellent example of the complexity of a technology that might be described simply as "weaving a basket." Yet this section describes only the preparation of the willow strips ultimately to be used for "weaving a basket."

Buffalobird-woman (vol. 20, 1916: 253–57)

This willow grows in the timbered bottomlands along the Missouri, and some trees are found in the timber groves out in the hills. It is not a plentiful tree.

We thought this willow a very useful tree. We cut the young trees to make bullboat frames and the frames of bark-covered carrying baskets. The wood was also used for making drums. We did not often use the wood for firewood for it did not give out much heat. Saddle stirrups were made of this willow but not saddle frames. *Maxoxica* wood is soft and was not much used for posts on that account.

Basket frames were made, as I have said, of *maxoxica* willow and also of the trunks of young diamond willows or *midahatsi popokci.*

But the chief use of this willow was to furnish bark from which our carrying baskets were woven. These baskets are usually made of two kinds of bark—boxelder and *maxoxica* willow.

The boxelder bark is a very light yellow, almost white, but the *maxoxica* willow bark was dyed black. This bark is yellowish-brown before being dyed.

I will now tell how the bark was prepared.

The bark was stripped from the trunks of young trees that were about 3 inches (75 cm) in diameter. If the trees were smaller than this, the bark would be too thin, and if larger, the bark would be too thick.

I cut the incision transversely across the trunk near the root and peeled the bark upward. As I did so, the outer bark fell away, leaving the inner bark. I peeled the bark thus all around the trunk in strips upward as high as I could reach. I then reached upward and cut it off with my axe or knife.

The bark should be taken when ripe which is just before June.

I first dried the bark on the corn stage with the tree or inner side up. When it had dried for about six days, I took it down to the river and put it in the water with a stone on it. When it was soaked, I brought it home and cut it into strips with a pair of scissors or a knife. I did not have to leave it in the river long. It soon got soaked through.

I cut the strips about as wide as the thongs around the handle of the broom I made for you. The wide strips came off the tree in lengths of about 7 feet (about 2 m), as that was the height one could reach. These long strips I cut in two, so that the narrow strips that I cut with the scissors were about 3.5 feet long.

When the strips were all cut out, I picked out the good ones, and going down to the creek or a pond, I filled a pail or pan with the mud that I found there. I added a little water until the mud was oozy and put into it as many of the strips as I wished to color. I left them in the mud six or seven days, and they became quite black. I then took them out and washed them in the river to get rid of the mud.

When the bark was being dried on the corn stage, the rain would not hurt it much even if it did wet the bark, but the boxelder bark had to be fetched in if a storm threatened or the rain would spoil it.

Maxoxica willow bark is tough and strong. Boxelder bark is softer and works easier. We used, and indeed now use, boxelder bark on the sides of the basket, but for the bottom of the basket we used the stronger willow bark, for which reason the bottom of a Hidatsa basket is usually black.

When I am weaving a basket, I am careful not to let the strips of bark get dry. If they begin to dry, I blow water over them with my mouth.

In the old days we often made baskets of *maxoxica* willow bark only, because this bark, being tougher, made the basket last a long time. Such baskets were uncolored and were a kind of yellowish-brown. They were rather pretty. I once made you a basket of boxelder bark all of one color. It is whiter than *maxoxica* bark. I never knew of boxelder bark being dyed black with mud.

I do not know of anything that was ever dyed black from the mud except *maxoxica* bark. Other women have tried to dye other bark in this way, but only *maxoxica* bark will become black.

A basket of willow and boxelder bark lasted a long time if it was used carefully and kept out of the rain, but if the owner loaded the basket with too heavy a weight or if she left it out in the rain, it soon spoiled.

Bark baskets were very useful. In old days we had no wagons and so carried our vegetables to the lodge in baskets. We gathered all kinds of cherries or berries in baskets, and we also used baskets for carrying stones and earth and ice.

I remember the baskets that were made when I was a little girl were still being used years afterwards.

FAMILY *Salicaceae*–willow family
GENUS *Salix* L.–willow
SPECIES *Salix amygdaloides* Anderss.–peachleaf willow

Sandbar willow (utilitarian, sweet)

HIDATSA NAME: *midahatsi hici* (willow red)
LOCAL ENGLISH NAME: red willow, river-bank willow (sandbar willow)
BOTANICAL NAME: *Salix interior* Rowlee

Despite Buffalobird-woman's assertion that only two kinds of willow are found on the reservation, she describes three, perhaps because she puts the peachleaf willow (*maxoxica*) in a slightly different category—it being a tree and used for baskets. The diamond willow is shrubby and an extremely important part of the roof of a lodge (forming a firm layer to support the grass thatch, which in turn supported the earth), and the red willow, also shrubby, was used for mats.

Buffalobird-woman (vol. 20, 1916: 293–94)

These willows were used for making mats. Such mats were not quite 3 feet in width and were in length about the distance across this room or about 8 feet. They were made of sticks about half an inch in diameter and sewed in three places with sinew. In old days we sat on mats on the ground. These mats were made like "lazy-back" mats and indeed were just like them excepting that the sticks were larger. Bits of cloth were inserted now and then between the sticks, just as I once told you of a "lazy-back." [The "lazy-back" was the legless equivalent of a chair among the Plains Indians.]

Sticks of this kind of willow were used in playing the game of *umakixeke*. [See an explanation of this game in chapter 10.]

The roots of the young plants in early spring before the leaves came out were very sweet and were chewed by children and young folks. After the leaves came out, the woody stem was chewed but not the roots.

The roots of these willows were also used in making a kind of round basket for playing dice.

Added by Wolf Chief:

Also on war parties, if out of food, we used to eat the roots of red willows, also those of the larger trees that stood on the edge of the Missouri bank, where the water had undermined the bank so that we could get to the roots. The roots we sometimes washed in water and sometimes cleaned with our hands. We chewed only the bark of the roots of the larger trees, but we chewed all the root-parts of the tender roots of young willows. We swallowed the bark as well as the juice. It was sweet and pleasant to the taste.

In old days also we used to gather rose pods, and we boys used to eat the rose blooms. They were fragrant, and we ate many of them.

Buffalobird-woman probably made the following remarks:

We now use these willows for making mats with which to floor our corn stages. Also for mats for making fish traps—willows about the size of a man's thumb for fish traps and a trifle larger for corn stages. The trees never grow large. I never saw one twice the size of my wrist. We think our name Hidatsa [Minataree] came from this willow.

Goodbird (vol. 16, 1914: 292–93)

My mother says that squash drying spits were made of native red willow, *midahatsi hici*. *Mida* means wood, and *hici* means reddish or pink. When the outer skin of one's finger, say, is peeled off, the color underneath we call *hica*.

This red willow is not what you call willow, however—not kinnikinnick (meaning red-osier dogwood–*Cornus sericea*).

We called those spits *kakuiptsa*, from *kakui*, squash, and *iptsa*, stringer or spit.

Small Ankle used to make these spits for us. He used to peel off the bark with his teeth. He would cut a lot of them and make a bunch of about three hundred and tie them together so that they would dry straight—would not warp, I mean, in drying.

Rods for making spits would be cut in June or early July, when the bark would peel off easily.

FAMILY *Salicaceae*–willow family
GENUS *Salix* L.–willow
SPECIES *Salix interior* Rowlee–sandbar willow

Heart-leaved willow (construction of lodges and fences)

HIDATSA NAME: *midahatsi pokpokci*
LOCAL ENGLISH NAME: diamond willow, heart-leaved willow
(Missouri River willow)
BOTANICAL NAME: *Salix eriocephala* Michx.

Diamond willow is not a precise name for a particular species because the characteristic diamond-shaped scar that forms when a branch dies is found on at least a half dozen species of willows (Lutz 1958), but the name is perhaps most commonly used for the Missouri River willow.

Buffalobird-woman (vol. 20, 1916: 291–92)

We Indians reckon that there are two willows on this reservation. Our word for willow is *midahatsi,* and our two varieties of willow are *hici,* or red, and *pokpokci.* I do not know what *pokpokci* means, for it is an old word and the meaning is probably lost.

Midahatsi pokpokci does not bend as flexibly as *maxoxica,* yet it bends quite well. So we used it for making sweatlodge frames. Also we used these willows for covering the roofs of our earthlodges. They were laid over the rafters, and when they were this kind of willow they lasted a long time. For covering the rafters we picked out willows that were about as thick as my thumb, young ones, using the whole plant.

Sometimes when we could not find *maxoxica,* we got diamond willows to make bullboat frames, but they were not so good. Only good ones, free of branches and knots, were used for boat frames.

Diamond willows grow thickly in the Missouri bottoms and about some springs. Diamond willows made good firewood, and we now use them in our stoves, for they make a quick hot fire. But we did not burn them in our earthlodges because sparks would fly out and burn people's clothing if they sat near.

37. Willow fence showing details of construction. Drawing by Goodbird.
(Courtesy of the Minnesota Historical Society)

We now use diamond willows for making posts. The bark is used for no purpose that I know of.

When we wanted to use diamond willows for covering the lodge roof, we cut them green and put them on the roof.

Diamond willows about 2 inches thick were used for making posts for fences around our garden. These posts were about 2.5 feet apart and were sunk in holes that were made by driving a stick into the ground and then withdrawing it. The willows made a very good fence. There were two rails on the fence, one at the top and another about halfway down the posts.

In spring the old posts were sometimes removed and new ones put in their place. The old ones were taken home for firewood.

Diamond willows were also used for making tobacco garden fences.

FAMILY *Salicaceae*–willow family
GENUS *Salix* L.–willow
SPECIES *Salix eriocephala* Michx.–Missouri River willow

Quaking aspen (tent poles, tool handles)

HIDATSA NAME: *matapuci aku makuhica* (*matapuci*, birch or poplar; *aku*, kind of; *makuhica* from *maku*, cottonwood; and *hica*, like. The birch that is like a cottonwood.)
LOCAL ENGLISH NAME: (not given) (quaking aspen)
BOTANICAL NAME: probably *Populus tremuloides* Michx.

Buffalobird-woman (vol. 20, 1916: 332)

This tree grows in the timber in the hills; some, not much, is found in the timber along the Missouri.

The trunks do not grow much over 9 inches in diameter. Being a very light wood these trees make good tent poles when a straight trunk of proper size is found; but as the tree tends to grow rather irregularly, favorable trunks are hard to find.

For tent poles the bark was peeled off the trunk as is the case for cottonwood poles. Some of the trees grow with a large foot that tapers rapidly upward with an upper trunk nearly of uniform size for some distance.

Good hoe handles were made from them, as they required a light wood. However, the handles for our iron hoes we usually made from cottonwood.

The bark of the tree is soft and was used for nothing by us.

FAMILY *Salicaceae*–willow family
GENUS *Populus* L.–cottonwood
SPECIES *Populus tremuloides* Michx.–quaking aspen

American elm (tea, toys)

HIDATSA NAME: *midai*
LOCAL ENGLISH NAME: elm (American elm)
BOTANICAL NAME: *Ulmus americana* L.

Buffalobird-woman (vol. 20, 1916: 300)

This tree grows in the Missouri River timber and less abundantly in the hills. It is not very abundant anywhere on this reservation.

We made tea of the inner bark after we got sugar.

Boys used to make pop guns of ash wood and shoot out of them chewed up wads of elm bark. The popgun made a cracking noise when it was shot.

The inner bark of the elm is tough, and we sometimes peeled the bark off and used it to bind those diamond willow fences we made about our gardens. We just used strips of bark for this purpose.

Elm wood was used also in making bows for hunting buffalos.

Goodbird:

Here is an Indian joke. Crow's Paunch said he once was off leading a war party when he saw ahead an elm tree with a big nest of hawk or eagle in it. He led his party thither, but when he came closer he saw that it was just one big wild turnip (*ahi*). It was so big that the party divided it among themselves and every man had enough to eat!

This is an old joke. The wild turnip does grow up something like an elm and has a blossom that looks like a bird's nest.

FAMILY *Ulmaceae*–elm family
GENUS *Ulmus* L.–elm
SPECIES *Ulmus americana* L.–American elm

Water birch (utilitarian—used in a pinch)

HIDATSA NAME: *matapuci aku cipica* (*matapuci*, birch or poplar; *aku*, kind of; *cipica*, black. The birch or poplar that is black.)
LOCAL ENGLISH NAME: (none given) (water birch)
BOTANICAL NAME: *Betula occidentalis* Hook.

There is not much to go on here, but this species does grow in northwestern North Dakota and has an almost black bark.

Buffalobird-woman (vol. 20, 1916: 320)

This tree was very little used by us. We made whip (quirt) handles of it. When we could not find *maxoxica* willow, we sometimes made bullboat frames of this wood. The wood was pliant, and the saplings if used for bullboat frames did not usually break in the making. The bark was peeled off to make the frames.

FAMILY *Betulaceae*–birch family
GENUS *Betula* L.–birch
SPECIES *Betula occidentalis* Hook.–water birch

Boxelder (sweet, utilitarian)

HIDATSA NAME: *mitetadiki*
LOCAL ENGLISH NAME: boxelder or ash-leaved maple
BOTANICAL NAME: *Acer negundo* L.

See "Plants used for utilitarian purposes."

FOURTEEN

Uses of
Wood

Gathering firewood

Gathering firewood was an activity, like hunting or hauling water, that never ceased. Each household used large quantities of wood just to cook meals and, in the winter, to provide some heat for the winter lodge as well. The two factors most heavily weighted in determining the location of a winter village (besides gaining shelter from the incessant winds of the plains) were the need for a reliable supply of water and a plentiful supply of wood.

With each family competing for a distinctly finite resource, gathering firewood became even more difficult. The longer people lived within a village, the farther they had to go to get wood for whatever purpose.

Women gathered firewood and frequently used the family dog with a travois to haul wood or to carry the bullboat on a travois up the river so that the firewood could be floated downstream to the village. Bullboats were fairly light, and some dogs could drag up to a hundred pounds, so burdening one of them with the bullboat for the outbound trip was not a problem. Only for snagging wood from the flooding Missouri River did the men become involved.

Wolf Chief (vol. 17, 1915: 47–52)

During the June rise of the Missouri River, we used to go down to the edge of the river to look for drift logs. At this time of the year there was always much driftwood coming down the current. The Missouri River rises when the snow melts in the spring and again about June. This latter event we called the "June rise." My father, Small Ankle, my mother, and myself would go down to get the driftwood. My father made a long hook by taking a long tent pole and lashing a wooden hook cut from the fork of a small

38. Buffalobird-woman with a load of firewood. Drawing by Goodbird; comments by Gilbert Wilson. (Courtesy of the Minnesota Historical Society)

ash tree to one end. With this he would wade into the water and catch any logs that came drifting near the shore.

The June rise was awaited eagerly by all the villagers, for they all hoped to catch drift logs for firewood. Small Ankle used to go out at sunrise just after breakfast, and we went with him. We used to work till we were hungry and at this point returned to the lodge. After the noon meal we went out again.

All the families of the village were down at the river on the same errand, and the boys of each household helped their parents as I helped mine.

It was the custom of us Hidatsas to take a bath in the Missouri every morning through the summer season, but on these days we did not think it necessary to take a bath, since we were working in the water all day anyway.

Wood suitable for fuel was scarce around Like-a-Fishhook Village. Most of the fuel wood in the timber along the Missouri already had been cut down by Indians. When this driftwood came with the June rise, our people were almost crazy about it, and everyone was eager to get all the driftwood that he could. I think this driftwood season used to last about ten days, then the river would begin to fall again. All our village worked to lay in a good stock of firewood while the rise lasted.

39. Small Ankle snagging driftwood from the Missouri River. Drawing by Goodbird. (Courtesy of the Minnesota Historical Society)

Small Ankle used to wade into the river up to his knees or up to his neck or, if the log were a good one and worth the trouble, he would swim out and get it.

Sometimes I would swim out and bring logs into the shore. I was just like a fish and could swim anywhere. I stripped off all my clothing so that I could swim without being bound with my clothes.

I was a good swimmer in those days, in any kind of weather. If the waves were high, I would turn over and swim on my side to keep the water out of my face and nose.

Besides catching logs by swimming out to them, I also had a wooden hook like my father's with which I caught drift sticks. Whenever I caught a good log that came floating by me, my father was glad and would call out, "Good, that is a good one!"

After the wood was dragged out of the water, it was piled up in piles on the shore nearby. Usually as my father hooked a log and brought it in to shore, my mother would seize it and drag it to the pile, but if it happened to be too heavy for her, my father caught one end and helped her with this labor.

The newly caught wood was let lie in the pile to dry for a little while. Perhaps that same afternoon or the next day, the women cut it into lengths—some of them four feet long—and packed it to the village. This fetching of the wood from the river to the lodge was always women's work. Small Ankle never helped at it, but my mother was helped by her dogs as were other women of the village.

My mother sometimes carried a log on her back if it was not too big, or she might carry four small logs or sticks. Some of these

40. Buffalobird-woman carrying a log back to the lodge. Drawing by Goodbird. (Courtesy of the Minnesota Historical Society)

sticks that she carried were four Indian fathoms in length. Sometimes she bore on her back a bundle of small sticks.

[An Indian fathom, as noted earlier, was the distance between
outstretched hands, roughly 5 to 6 feet (1.5–1.8 m).]

I have told you before how our tribe used the dogs with the
travois. Women often brought dogs down the river to help fetch
firewood to the lodge. At the time of the June rise, all the dogs'
travois were loaded only with shorter sticks, say, three feet in
length or less. There was no exact measure that we used in cutting these lengths, but very long sticks were never loaded on a
travois because they were likely to catch on the posts of the drying stages when the dogs were led through the village.

The dogs' travois were loaded down at the edge of the water,
but the dogs climbed the banks very well because the travois
were loaded very lightly. The woman who owned the dogs led
the way and the dogs followed. The woman also bore a load,
either a small log or two or three smaller branches.

Having arrived at the corn drying stage that stood before the
lodge, the woman would drop her own load under the stage, but
when she unloaded the dog's travois, she threw the sticks up on

the stage floor. This was not hard to do because the floor was not high. I have often helped my mother do this.

The reason for this was that the travois was loaded with small sticks. The larger logs which were brought up on the woman's back were put on the ground under the stage. The small sticks were thrown up on the stage floor to dry, and the larger ones were left to lie under the stage for the same reason—to dry them.

Wood that had already been dried was sometimes tossed up on the floor of the stage in the same way. I think this was done lest somebody might come and carry it away. I have heard that a lazy woman would sometimes, when out of firewood, steal some fuel from another pile, but I have never known of this being done in my own family experience. Thus, if wood were tossed up on a stage floor and the ladder taken down, it was safe from anyone inclined to steal.

When wood was desired for the lodge fire, it was taken from either the pile on the stage floor or the pile on the ground beneath. The woman would mount the ladder, throw down some sticks to the ground, and descend. She then would carry the fuel into the lodge in her arms just as whitemen would carry it.

Sometimes wood was also piled on the roof of the entryway that covered the entrance to every earthlodge. This might be dry wood or might be wood newly taken from the river in the June rise, but the drying stage and the entryway roof were the only places where the firewood was piled outside of the earthlodge.

After the June rise, when the river had fallen again, much driftwood was to be found deposited on the sandbars of the river. This driftwood was likewise borne to the lodge in the same way as the wood rescued from the current.

Wolf Chief (vol. 17, 1915: 53)

Storing wood in the lodge

In the earthlodge, wood was piled in the first two sections back of the corral. The sticks were laid horizontally, or lengthwise with the wall, and they were prevented from rolling inward by a stake driven in the ground.

[As noted previously, the corral was the stall constructed usually to the right of the entryway of the lodge where the most valuable horse or horses would be sheltered from enemies or harsh weather.]

In summer the pile of wood here was small, merely a supply kept against rain in threatening weather. In our winter lodges down in the timber, the piles were much larger. In a winter lodge, wood was not kept outside of the lodge but always brought within.

When a woman ran out of wood and her fire needed replenishing, she went out, fetched a load, and put it between the couch and the fire. She just dropped it on the floor loosely, taking no trouble to pile it up in any ordered way.

Digging-sticks

I think that there is the distinct possibility that the digging-stick was the first tool used by the predecessors of modern humans, but unlike stone tools, digging-sticks do not persist in the archaeological record. It is as close to a universal tool as possible. The Hidatsas used large digging-sticks for excavating postholes for an earthlodge, medium-sized digging-sticks (4 feet [1.3 m] or longer) for gardening, and shorter digging-sticks for excavating eagle-trapping pits. My wife and I use them every year when planting our Hidatsa garden, and they are perfect for making the hole into which we place a kernel of corn. By the time Buffalobird-woman was gardening, iron tools had replaced the bison scapula hoes (she never used one), but the 4-foot digging-stick was too handy to give up. Besides its utility for loosening soil, the length of the digging-stick was the same as the distance between corn hills and therefore a handy unit of measure [we find]. Furthermore, digging-sticks were the ideal tool for digging prairie turnips (*Pediomelum esculentum*), which remain a treat. Braids of prairie turnips are still sold at Indian gatherings in the Northern Plains, and they may still be dug from the prairie with a digging-stick.

Gilbert Wilson summarizes a conversation with Buffalobird-woman (vol. 8, 1909: 93–94)

Digging-sticks were called *makipi*. They were made of ash. Buffalo fat was rubbed into the wood, and it was held over a slow fire. This made the wood at the point firm and almost as hard as steel. Very often the instrument was thus exposed over the fire for its whole length, although only the lower third so treated was enough to make the instrument serviceable.

41. The basic garden or all-purpose ash digging-stick. Drawing by Frederick N. Wilson. (Courtesy of the Minnesota Historical Society)

With this simple instrument, corn was planted and holes were dug. Prairie turnips were also dug with it.

It was a woman's implement. When using it, the woman folded her blanket into a pad and rested it over her abdomen. Against this the handle of the digging-stick was placed, and the digger rose on her toes and forced the point of the digging-stick into the soil with a downward and partly rotary or swinging motion of her body. For prairie turnips, this was very effective as the writer can assert from experience. Buffalobird-woman says that for stony ground it was superior to the spade.

Digging-sticks for digging deep holes such as the postholes of an earthlodge were made longer and somewhat heavier, but the type was the same. A digging-stick would last a great many years.

Buffalobird-woman (vol. 11, 1912: 5)

We always had a digging-stick in our lodge. A good digging-stick had the sharp or digging end cut from the root end of the young tree from which it was trimmed. The root end of a young tree would be largely free of branches and easier to cut to shape and would also be stronger.

Digging-sticks were also used to dig up wild turnips. When one was on horseback looking for turnips, she carried her stick athwart the saddle in front of her. Her saddlebags on either side of her horse were loaded with turnips.

A digging-stick was a very useful instrument besides its use in the garden and for digging turnips. It was used for making tent pole holes, postholes, or in fact any kind of hole, and it was used as a crowbar for rolling logs.

Wolf Chief (vol. 17, pt. 1, 1915: 129–30)

Temporary digging-sticks made on the eagle hunt

As soon as we were through (with breakfast) we took the hoe and knives and some of the digging-sticks we had made the evening before and started out.

The digging-sticks were made of ash, of dry wood, and not green. We had made three of them about as long as a man's forearm and about as thick as his wrist. These we shaved to a point and covered with deer fat and then held them over the coals to harden the point. The digging-stick was shaved to one side [to make a flat surface]. When the point got dull or worn out, we would hunt up a big boulder and rub the point on it to sharpen it again.

Mrs. Packs Wolf (vol. 10, 1911: 263–64)

From Wilson's description of items in a medicine bundle

XI. Small digging-stick. In old times, the digging-stick was used to dig out the eagle hunter's pit. There was just one singer who went out and sang his sacred songs for he was the leader of the party. He would sing and mark out the corners with the digging-stick, and the pit would be dug afterwards with an iron hoe. We kept up this ceremonial use of the digging-stick because it came down to us from our fathers. In old times the whole pit was dug with a digging-stick.

Mortar and pestle

Buffalobird-woman (vol. 11, 1912: 311–12)

[The following is Buffalobird-woman's elaboration on an item Wilson had purchased from her for the American Museum of Natural History: $7.00 of Buffalobird-woman.]

When I was about twenty years of age I went down one day in spring before the Missouri had broken and crossed the ice to a sandbar. There I found an ash trunk that had floated down probably the previous summer and lodged on the bar. With an ax I cut off a section of the trunk just above the roots.

I cut the section five inches longer than the mortar is now, and I carried it home with my shoulder strap.

My father set the section of root-end up near the fire in the lodge and prepared to burn out a cavity in it. For this we always used cottonwood coals from the fire because they burned with a fierce heat. My father began at once because ash burns well even when green.

We had two ways of lifting coals from the fire—with a split piece of firewood which was flat like a board or with a split of kinnikinnick stick used like a pair of tongs.

My father took up a live coal and laid it on the middle of the root-end surface of the section of trunk. With a hollow reed two feet long he blew on the coal. As it burned up, the ash blew away, and it got smaller and smaller until my father brought out another coal and put it in its place. He kept the fire from creeping too near the edge by dipping his finger in a wooden bowl of water and wetting where necessary.

This coal that my father put on the end of the log section was about the size of a [silver] dollar.

He blew, working thus most of the day until his blowing fetched a headache so he stopped.

When the cavity was about four inches deep he began to hollow it out inside. If it threatened to burn too far into the side my father would take a stick a foot long with a bit of cloth tied to the end, dip this cloth end in the water, and check the fire.

As the cavity went deeper it grew quite hot within. To prevent the cracking of the wood, my father anointed the upper part inside with buffalo fat.

The work continued three or four days when the cavity was about 8 inches deep. (The cavity is now 13 inches deep, made so by long use.)

When the cavity was all burned out, my father cut the mortar to the shape as you see it now.

We had an ash pestle in the house which we used in this mortar until it wore out. It had been in use a long time before. My father went into the woods and made another, the one I sell you now. He made it about forty years ago. It has worn down 7 or 8 inches by use.

The weighted end of the pestle, you note, is ax-marked. In the lodge we habitually used to cut meat or a bone or the like with an ax, on the weighted end, as on a block. Hence the ax marks usually found on a pestle.

The first whiteman we entertained in our house ate meal pounded in this mortar.

This mortar had been used in our family for nearly fifty years. The last time it was used was four years ago; since then I have had it by me to remind me of old times, but now I am willing to sell it to you if I get a big price.

(The ordinary price for mortar and pestle is about $2 or $3. This specimen was bought partly as a concession to Maxidiwiac, Buffalobird-woman. GLW)

Making a bullboat frame

The circular bullboat is the most expedient solution to making a portable boat from a single bison hide. It is very much like the Welsh coracle in appearance and was one of the reasons that various early travelers were convinced that the Mandans (with whom the Hidatsas shared an essentially identical material inventory) were a lost tribe of Welsh (e.g., Catlin 1973: 2:259). A bison hide was stretched over a constructed frame, hair side out. The tail was left on and was literally the stern of the round boat. A bullboat was light and fairly portable, especially when pulled by dogs upstream on A-framed travois. Bullboats were not meant to be paddled against the current.

Once used, the bullboat was dried by placing it on the earthlodge roof, where it would also serve as a cover for the smoke hole (or the "sunbeam," as the Hidatsas called it) when it rained. A bullboat frame might also be placed over the smoke hole/skylight to keep dogs and children from falling through. Paddling was a bit tricky and was done from the front. This made the boat tip forward so that a weight of some sort was needed at the stern. It was not a stable craft, and when the river was rough, so was riding in the bullboat. As a newborn, Goodbird was flipped into the Missouri River when the craft in which he was riding was tossed about and his cousin lost his grip on the baby. Thanks to the buoyancy of Goodbird's wrappings, he was quickly rescued.

Buffalobird-woman (vol. 25, 1912: frame 0082)

A bullboat frame should be of *maxoxica* or basket-bark willows [*Salix amygdaloides*] if possible. Next in preference is ash. When out on a hunt, our people sometimes could not get *maxoxica* willows or ash and would use chokecherry or red poplar.

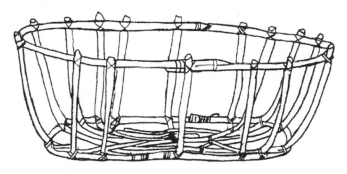

42. A bullboat frame, which is about 5 feet (1.5 m) in diameter at the top. Drawing by Goodbird. (Courtesy of the Minnesota Historical Society)

Our name for this red poplar is *matapuxi aku cipica* from *matapuci*, poplar; *aku*, kind of or sort of; *cipica*, black or dark [*Betula occidentalis*–water birch]. There is another tree we call *matapuci aku makukike*—cottonwood or red poplar of the kind like cottonwood. *Matapuci aku cipica* grows a little larger than a chokecherry tree.

(The author was unable to verify these two trees that Goodbird calls poplar. GLW)

[Although one has been described as *matapuxi aku cipica*–water birch, *Betula occidentalis*, the other is unknown.]

Gilbert Livingston Wilson (vol. 25, 1911: frames 0071–0073)

Further observations concerning bullboats

July 18th 1911 the author borrowed a bullboat of Mrs. Shultis at Independence and with Goodbird and the latter's wife rowed or floated down the Missouri to Hairy Coat's lodge. Goodbird's wife rowed to show how the boat was managed in old times, usually by women. The author obtained a number of good photographs, landing on a sand bar for the purpose.

Goodbird called attention to the fact that a good boat was built with the first or lower ribs of the frame laid parallel with the spine or head-to-tail line of the covering hide. This is done to secure the result that these ribs will lie with the hair. In paddling the boat, the rower kneels in the forepart of the boat, or part where the neck of the hide lies. Such a position, with the rower dipping the paddle directly before her, prevents the resistance which in any other position would cause the water to make by ruffling up the hair which is always on the outside of the covering hide. Resistance is also avoided, which the ribs swelling out the tight drawn hide would certainly make if the boat were paddled in any direction not parallel with the way the lower ribs of the boat lie.

On our return, by oversight, the boat was not put away under cover and for a day or two stood by the ice house. Not, it is true, in a hot sun but exposed to a rather strong, dry prairie wind. The wet hide cover dried too rapidly and shrank the frame, breaking one of the ribs and splitting the hide covering itself. Commenting on this, Wolf Chief said that both accidents were usual and could be mended—the rib by bracing it with a section of the limb of some wood and the rest by sewing on a piece of softer, dressed hide after first wetting both.

During the journey to Hairy Coat's lodge Goodbird made the following observations:

A bullboat ought to be made as flat bottomed as possible. This bullboat is not a very good one, for it is rather round bottomed and such a boat is easily upset.

A bullboat is usually paddled by one person kneeling (or sometimes sitting) in the forward part of the boat and dipping the paddle directly before. In old times the bullboat was rather a woman's craft, although men also used it. Often a war party would float in a bullboat by night down into the enemy's country, steal horses, and ride back after abandoning the boat.

A bullboat is sometimes paddled by two persons, one on either side as in a canoe.

A bullboat should be built so that the covering hide runs lengthwise with the bottommost ribs of the frame. This avoids friction against the water by the ribs.

The boat should be paddled so that the tail of the hide is behind, that the hair may not rise and make friction against the water.

The hide should always be laid hair out and should be put on green from the animal.

If there is a leak, thrust a little twig through the hole and plaster on the inside with thick, sticky mud. If the leak is small, the twig plug is omitted, but mud is plastered over the aperture.

When a boat has but one occupant, it is balanced by a heavy stone (or a large chunk of the dried, hard clay common along the Missouri).

A bullboat is better than a (whiteman's) skiff when the river is running full of ice (as at the spring break up).

Making a wooden bowl

Although a centuries-long tradition was to cook food in ceramic vessels, in Buffalobird-woman's time it was cooked in brass and iron pots. Food was consumed from wooden bowls, a practice which continued through the nineteenth century. In the days before iron tools became common, the labor required to manufacture a wooden bowl was considerable, and even with iron tools each bowl required a significant investment of time. Wooden bowls, of course, did not break easily and lasted for years. Women made pottery, and men made wooden bowls.

Wolf Chief (vol. 27, 1915: 65–67)

I once saw my father make a wooden bowl such as we formerly used. My tribe was in winter camp, just opposite Independence, and the weather was quite cold. I was about nine years old [ca. 1858].

The bowl was wrought from the knot [burl] of a cottonwood trunk. A kind of sore forms on the trunk and leaves such a knot when it heals. As he told me afterwards, my father felled the tree and cut out the knot with an ax. He fetched the knot home about three o'clock in the afternoon. It was about 18 inches (45 cm) in diameter and 6 or 7 inches (about 16 cm) thick.

Using a big butcher knife with a chopping motion my father worked the knot into a round shape. To hollow out the inside he used an axe with a short handle.

I was a boy at the time and did not pay much attention to my father's work nor closely notice all he did. I do not think he first dried the knot but began working it at once. I remember he used the butcher knife to carve out the inside, working slowly and shaving carefully.

Sometimes I saw him using a kind of red stone quite light, which we find along the Missouri. He used this I think to finish off the bowl and smooth it.

[This is locally and erroneously called *scoria*; geologists use the term *clinker*. Its red color makes it a prominent feature in the badlands just to the west. Real scoria is formed from frothy lava. The local scoria is the product of burning lignite veins. Lignite is a soft coal found in strata of varying depths in the sediments throughout the area. When ignited by spontaneous combustion, a lightning strike, or a prairie fire, it heats the overlying and underlying clay, baking it to the color and consistency of brick. On occasion the clay is heated to the melting and even boiling point and may flow and froth like lava. Pieces of this stone can often be found floating in the Missouri River and washed up on the shores. Chunks of scoria are abundant in the badlands. Some are rough, full of holes, and like pumice and were widely used for smoothing wood.]

One day I saw him rub the bowl with ghost whistle rushes [scouring rushes–*Equisetum hymale*] to polish it.

How many days my father worked on the bowl I do not know. It was a fine, big bowl when finished and completely round; that is, it had no handle or ear as you call it.

Before my father made this one, there were five bowls in our family's possession. The one newly finished made a sixth. Some of these older bowls were rather large.

The largest was shaped somewhat like a boat and was about 20 inches long by 17 inches wide. It had a dog's head carved at one end and a round knob at the other. A second bowl was round without head or knob and about 13 inches in diameter. A third bowl, also round, was 12 or 13 inches in diameter as were the two remaining bowls. The sixth and newly finished bowl was larger than these last but not as large as the one that was shaped like a boat. I presume my father made all these bowls, but I do not know for sure.

The family bowls were kept in two places: in a skin basket hung up on one of the lodge beams or in a bag that swung from

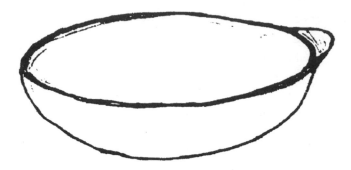

43. A large wooden bowl with a lug or handle. Drawing by Goodbird. (Courtesy of the Minnesota Historical Society)

one of the rafters. Sometimes I saw these under my father's bed near the fireplace. I do not remember whether the skin basket hung from a thong. I was rather small and did not observe closely.

I have heard that such wooden bowls were made and used since very old times. My father once told me, "I have heard how they made these bowls in old times before axes and knives of iron were known in our tribe. They were hollowed out with coals of fire. The maker blew and blew on these coals, then rubbed the bowl into shape with rough stones which he kept by him. The man blew the coals with his mouth, not through a reed, at least so far as I know."

This, I think, is all the knowledge I have of making a wooden bowl.

Wolf Chief (vol. 17, 1915: 32–33)

I have already spoken of the wooden bowls that my father and I ate our food from when we were on the eagle hunt. The bowl which my father and I had on the eagle hunt was without a handle as indeed were most of the bowls. Still, other bowls with handles were common enough in the tribe.

These wooden bowls we used to make from cottonwood and boxelder. They were made from nubs or knots [burls] that we found on the trunks of trees. Here is a sketch of a knot on a tree that stands near my corn field.

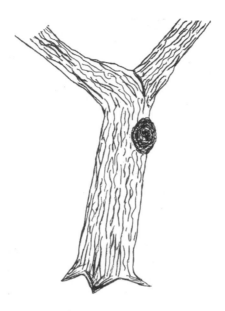

44. A burl on a cottonwood tree. Drawing by Goodbird. (Courtesy of the Minnesota Historical Society)

These knots were much harder to find on boxelder trunks than on cottonwood.

Wooden bowls were quite valuable as they could be carried along on journeys without any danger of their getting broken.

Wooden bowls were carried on a journey wrapped up in a piece of old tent-skin and usually borne on the back of a horse. When the horse was packed with his load, the bowl was the last thing to be put on so that it would have nothing on top of it to break it.

When I was about sixteen years of age I once saw a wooden cup or drinking bowl such as were once not uncommon in my tribe, so I have heard. It was about the circumference of a cup, but only an inch or a little more deep. A hole was drilled near the edge and the cup was tied to the belt of the wearer just over the right hip. This cup was worn by Cherry In Mouth. Such a cup was useful to a man on a war party to drink from. A thong went through the hole to tie the cup to Cherry In Mouth's belt.

45. A wooden cup. Drawing by Goodbird. (Courtesy of the Minnesota Historical Society)

This is the only wooden cup of this kind that I ever saw in my tribe, but I have heard of their use in old times from others.

Buffalobird-woman (vol. 20, 1916: 155)

Wooden bowls were usually made from cottonwood and box-elder. I do not think my tribe made them from ash, but I think some other tribes did.

For a bowl, the big knot should be cut out in the fall. If it was cut in the spring, the wood would crack, but if in the fall the wood would never do so.

An old woman in our village named Cedar Woman and some-times called Sage, an Amahami, used to make wooden bowls.

[The Amahamis were one of the three Hidatsa groups living at or near the mouth of the Knife River. The others were the Awatixas and the Hidatsas.]

Boxelder trunks are sometimes quite large, 15 inches (38 cm) or more in diameter. Cedar Woman would cut out a section of the trunk that was yellow inside—that is, there was no red in the wood. She would split this down the middle and begin to cut out the center with a big knife. When she had it as deep as she could, she scraped it out further with a piece of steel with a bit of skin wrapped around it for a handle. But I feel sure she shaped the bowl first outside and then hollowed it. She used to sell many of these wooden bowls.

Rakes (and the bison scapula hoe)

A garden in the spring is a mess. There are cornstalks, squash vines, and other dead vegetation that need to be cleared. Having rakes for the task greatly simplifies the gathering of remnants of the previous year's garden and then carrying them off to a site to be burned. Note that Wolf Chief expresses a preference for rakes made of ash, but his sister thought otherwise—that rakes made of antlers were decidedly preferable and were less of a risk to both garden and gardener.

Once the garden had been cleaned, it could be prepared for planting and subsequently for weeding and hilling, which required the use of bison shoulder blade hoes.

Buffalobird-woman (vol. 8, 1909: 92)

Rakes were made of ash. Five long withes or sticks of ash were bound together, the nether ends of the withe being spread out and bent over into teeth—much like the spread and crooked fingers of the human hand when raking, say, through a pan of sand. Wooden rakes were open to one objection. They were believed to be the abode of worms which might enter the body

46. A bison shoulder blade (scapula) hoe, the basic garden tool used by horticulturalists throughout the Great Plains. Drawing by Frederick N. Wilson. (Courtesy of the Minnesota Historical Society)

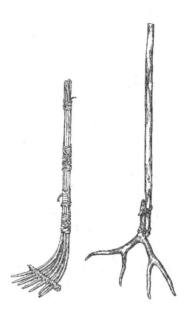

47. A wooden ash rake and an antler rake. Drawing by Frederick N. Wilson. (Courtesy of the Minnesota Historical Society)

For this reason wooden rakes were not thought to be a good instrument for preparing the bed of the field for corn; the worms that are often found on young corn were firmly believed to have escaped out of the wood of the rake used in preparing the ground. Rakes made of deer horn were therefore preferred. The antlers of a black-tailed deer were best, as they were "full of branches."

Wolf Chief (vol. 9, 1910: 291–94)

We made wooden rakes from ash. The example that I have made for the museum is made with the top piece of a forked branch, and the other teeth are bound to this forked piece. This is the best way to make a rake.

A wooden rake should have six or seven teeth. A stick is shaved down to proper size at one end while still green and held over the fire with the end bent at the proper angle to make the teeth required.

I accomplish this as follows. When the stick is well heated, I bend a tooth into shape and tie it with a rawhide thong. When all are treated thus, they are gathered into a bundle and held over a fire until dried. As each stick seems dried sufficiently, it is laid down out of the bundle.

When all are dried, the rawhide strings are taken off.

The various sticks are then gathered up, and a small section of a stick is split for the brace that goes transversely across the teeth of the rake to hold them in place. The two teeth of the forked piece are bound in place first. Next are bound the teeth that lie on either side. Rawhide thongs are used to bind these in place. Last of all the handle is bound about.

The first that rakes are mentioned in my tribe, that I know anything of, is in the story of the Grandson. There is a little lake down near Short River and an old woman lived there—Everlasting Grandmother. There is a flat piece of ground there 5 miles long by 1.5 miles wide. All that flat land was Everlasting Grandmother's garden. Her servants were the deer that thronged the nearby timber. These deer worked her garden for her. All buck deer have antlers, and with their antlers the deer raked up the weeds and refuse of Everlasting Grandmother's garden.

Now deer shed their antlers. Everlasting Grandmother got these shed antlers and put them on sticks, and so we got our first rakes. Her grandson saw what she did and afterwards taught the people how to make rakes also.

In later times we came to make rakes of ash wood instead of deer antlers. But we still reckon the teeth to mean the tines on a deer's antler. Sometimes deer have six, sometimes seven tines on an antler. So we made our ash rakes, some with six, some with seven teeth.

If the grandson had not seen what his grandmother did, we should never have known how to make rakes, either of antlers or of ash.

In our tribe in old times, some men helped their wives in their gardens. Others did not. Those who did not help their wives talked against those who did and said, "That man's wife makes him her servant."

And the others retorted. "Look, that man puts all hard work on his wife." Men were different. Some did not like to work in the garden and cared for nothing but to go around or be off on a hunt.

My father, Small Ankle, liked to garden and often helped his wives. He told me that that was the best to do. "Whatever you do, help your wives in all things," he said. He taught me to clean the garden, to help gather corn, to hoe and to rake.

My father said, "That man lives best who helps his wife."

Observation by G. L. Wilson

The writer asked Wolf Chief to demonstrate the use of the rake, which he did. The rake was drawn with a slightly rotary motion, not drawn straight in with the force laid on all the teeth at once as we do with our iron rakes. "The wooden teeth," Wolf Chief said, "will not stand the strain." [I think what Wilson means is that the rake was used with a sweeping motion.]

Paddle for working clay pots (cottonwood bark)

The process of making ceramic vessels as described here is familiar to archaeologists. The technique, called "paddle and anvil," can be found dispersed over much of North America. A vessel is formed by hand from a mass of clay until it has the shape roughly desired for the finished vessel. As the mass of clay is worked, the smooth anvil stone is held against the interior of the pot and the "patter" or paddle is "patted" against the clay directly over the anvil. This compacts the clay, and the rotation of the vessel (by hand—no wheel was used) and the continual paddling and manipulation of the clay creates the final form of the vessel and determines the texture of the clay. The ridges or design on the paddle impressed on the still-pliant clay can be completely obliterated by rubbing the vessel with a smooth stone or only partially smoothed over, as is commonly the case.

Gilbert Wilson (vol. 9, 1910: 8)

From an inventory of items purchased for the American Museum of Natural History

(30) $.25 of Sioux Woman. Pottery tool. Mandan name is *medahitki*–pottery patter; I shall call it a patter. Was used by Hides and Eats, Mandan.

"The pot, when formed and still damp, was set on the ground. A smooth flat stone was held inside and the sides and bottom were tapped over with the patter, care being taken to make the tapping strike against the stone held within. This made the pot firm. When pot was dry it was fired with cottonwood bark of which also the patter is made. The pot was turned upside down on a pile of bark and bark was heaped over it. Bark was fired and pot heated red hot.

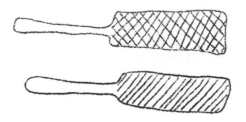

48. The "pottery patter" made of pieces of cottonwood bark. Drawing by Goodbird. (Courtesy of the Minnesota Historical Society)

A pot when of good clay would stand the firing (or cooking as she called it) without pieces being flaked off by the explosions which always attack poor clay in the firing process.

The effect of the serrations on the patter is to leave marks all over the pot. These are smoothed off or nearly so with a smooth stone. When pot is dry, these serration marks may be left in, to add to the pattern. Otherwise the pot is polished with a stone, as just explained. When pot is fired, it is given a final polishing with fat."

Butterfly adds, "Such a pot will hold water all summer. In an iron vessel the water soon dries up.

FIFTEEN

Arrows

Significance and utility

Even during the period covered here, roughly between 1850 and 1900, bows and arrows were still very much part of Hidatsa culture. Making flint arrowheads and stone axes became a lost art almost immediately when iron arrow points, knives, and axes became widely available in the first decades of the nineteenth century, just as the making of clay pots ceased almost immediately when brass and then iron pots became available at about the same time. But, particularly in the early through mid-decades of the nineteenth century, bows and arrows were less expensive, easier to obtain and maintain, and more reliable than the firearms obtained from traders and through trading with other tribes.

The acquisition of better firearms, particularly after the Civil War, did not, however, eliminate bows and arrows from boys' play. Both bows and blunt arrows were fairly easy to make or obtain from obliging relatives, and arrows made of grass (big bluestem) were readily available from midsummer on. Battles between groups of boys using grass arrows were a common, if not hazardous, pastime. Boys hunted small birds with their bows and blunt arrows, thereby developing skills highly valued by people who relied heavily on hunting for subsistence.

Wilson recorded a great deal of detail on the manufacture of both bows and arrows, but more information on arrows is included here because the details of making bows are largely details of woodworking, whereas arrows, besides having diverse uses and forms, were of far greater symbolic significance.

Adapozis or Burnt Arrow (or Flaming Arrow) was one of the Thunderbirds who, when he saw what a desirable place Earth was, turned himself into an arrow and launched himself from the skies to take residence here. After the destruction of his first village by enemies, he founded a new village wherein each family was formed from a different part of an arrow—the arrow that was Adapozis.

Wolf Chief (vol. 7, 1908: 28–29)

"Now that their enemies were all dead, Burnt Arrow thought to have another village to replace that which had been destroyed. He therefore said:

"Behold, I have my arrow and out of my body, the arrow, I will make my village. And I will make it thus: my body—the shaft of the arrow—shall be one earthlodge and one family living therein; the three feathers on my body shall be three earthlodges and three families; the nether sinew that binds the feathers on my body shall be one earthlodge and one family; the upper sinew shall also be an earthlodge and a family; the three wavy lines on the sides of the shaft shall be three earthlodges and three families, the nock shall be one earthlodge and one family. The arrowhead shall be one earthlodge and one family, and the sinew tying the arrowhead shall be one earthlodge and one family. In all there were twelve earthlodges and twelve families."

Making arrows

Wolf Chief (vol. 10, 1911: 41–55)

We usually made our arrow shafts from juneberry [*Amelanchier alnifolia*] shoots. We also used snakewood [*Gutierrezia sarothrae*] and split ash. Ash [*Fraxinus pennsylvanica*] made excellent war shafts, being tough and strong and seldom breaking. Snake wood was used only in war. It grows in the vicinity of the Little Missouri River [in the badlands]. It grows on the sides of clay hills. It has sharp thorns like rose bushes, somewhat.

We considered snakewood to be a magic wood. It was used usually only in war, although I have shot rabbits with snakewood arrows. Snakewood was very dangerous, because it was poisonous. It grows in sticks about the size of one's finger and turns yellow when the bark is removed. If a splinter wounded one while working the wood, the wound would swell up and one might die.

Once a Montana Indian by accident had a splinter of snakewood run into his thigh. The wound swelled up, and he died the next day. He had much pain and cried and groaned.

Snakewood arrows were brought to an elongated point. This point was expected to break off in the body of an enemy and poison him. In very early times, we hear that our tribe fought with ash-wood arrows and snakewood without heads. I have already described the snakewood arrows. The arrows of ash-wood, I have heard, were cut into something the shape of a broad head but not so broad. I have made one arrow of the shape that I imagine it must have been from the traditions I have heard concerning it. The arrows were oiled with bone-grease and browned, but not burned black over the tip.

Later our tribe found that juneberry made good arrows. The

wood was firm and strong, and they made arrows from it and tipped them with flint heads.

All this I hear only by tradition, and I tell it as best I can. In my day we had plenty of iron, and we no longer used flint or wooden headed arrows.

Juneberry shafts may be considered the usual and typical arrows of our tribe. Adapozis himself was a juneberry arrow and magic. We learned to make juneberry shafts from him.

When fresh shafts were cut, they were first debarked with a knife. These shafts were cut four or five inches longer than the arrows were expected to be. Ten of these shafts we bound into a bundle with four cords tied around in different places. This was to hold the shafts straight. This bundle was hung up in the smoke hole for a couple of days.

When thoroughly dry, the cords were taken off the shafts. Each shaft was now put through the hole of a bone polisher, made from the [spine of a] vertebra from a buffalo's neck [the hump]. The shaft was twisted about and worked around in the hole until the wood was made firm and solid. These polishers were sometimes made from buffalo ribs. [Archaeologists usually refer to these tools as "shaft wrenches."]

When the shaft came from the polisher, it was creased and rough. Two pieces of cottonwood bark were now chosen, and a groove cut in each so that they would fit together and just let the arrow pass through the groove. The groove was coated with glue. Pounded flint was poured over the glue and let dry. This flint was pounded with a quartz stone upon solid rock. The shaft was drawn through these two pieces of bark, and the pounded flint acted like sandpaper.

The shaft, thus smoothed, was given another polishing with native sandpaper. A piece of tent cover eight inches square was coated with glue, and powdered flint was poured over it. The arrow shaft was held with one end enveloped in this sandpaper in the left hand and rolled with the right on the right knee with an alternative rotary motion, thus giving a fine polish. A good arrow could not be made in a hurry. [Traditionally people used two pieces of sandstone, or "scoria," from the badlands, which, when placed together, looked somewhat like a hotdog bun. A groove the size of an arrow was carved longitudinally down the center of the two stones, which were then held together with the arrow in the center. Moving the "arrow shaft smoother" from one end of the shaft to the other removed any roughness. The methods described by Wolf Chief sound as though they may have been an adaptation of sandpaper observed at the fort adjacent to Like-a-Fishhook Village, Fort Berthold.]

Three wavy marks were traced down the shaft. I have heard that it was done with a sharp flint held pressed against the first and second fingers by the thumb, the shaft being drawn between the two fingers with a rocking motion. These marks extended from the feathers nearly to the head.

Burnt Arrow, or Adapozis, used these three lines and called them lightning. We make them likewise because he taught us.

The notch was cut for the string. In my day we took an old butcher knife with shallow teeth like a saw and used it to cut out the notch. How it was done in older times I am not sure, perhaps with flint, or as you suggest with a bone saw. I do not know.

I see I have omitted to say that at the end of the first day, the drying bundle of shafts was taken down from the smoke hole and the shafts oiled so that they might not crack. We did not oil them a second time.

The iron heads were put on. The shaft was split with the saw-knife for a little way. Glue was smeared on the base of the head, and it was bound in place with wet sinew.

The feathers for the arrows would have been gotten ready beforehand. We split the feathers and scraped the split surface smooth. In our day we did this scraping of feathers of small birds with a knife, holding one end of the split feather in the mouth. But we did not do so in old times when eagles' feathers were usually used because we thought eagle feathers were magic and might poison us if we put them in our mouths.

This is the method formerly used for eagle feathers. A small round stick was inserted under the toe of one's moccasin with the end projecting a little. The quill of the eagle feather was inserted under this stick and pressed down to hold it firm. The surface was scraped or, better, ground smooth with a piece of the scoria or natural brick that floats down the Missouri River. This stone or brick is red.

When a large number of feathers were prepared, they were put away in a feather case with the glue stick and instruments for arrow making. All these were now brought out, the case opened, and the feathers laid conveniently by, ready to be put on the shafts.

Three feathers were used on an adult's arrow. They were first bound to the shaft at the ends nearest the nock—the upper end of the arrow, I will call it. A bit of sinew was used. It was not thought necessary to use glue. The sinew was merely wetted. Indeed we did not think it wise to use glue, as a jagged end or projection might rise in the glue spot and hurt the hand as the arrow was discharged. The sinew was wetted and drawn through the teeth to make it thread-like. One end of the sinew was held in the teeth, and the arrow revolved in the two hands, whipping on the sinew. The thumb was passed around the now whipped sinew, pressing it very flat and even. This was the most delicate part.

The feathers were now bound at the lower ends in exactly the same way. Each arrow, as the binding of the feather was finished, was leaned up against a small log before the fire, to dry the sinew. When quite a number had been thus made ready, the arrows were ready to glue.

The glue-stick was lifted from the feather case and a small pot of water set handily by. The glue was dipped in the water a moment, a few coals were raked from the fire, and the new-wetted glue-stick was held over them and revolved over the fire about four times. The glue-stick now had a soft, thin coating of glue on the outside of the glue-mass.

The arrow maker then takes up a feathered shaft and holds it in the left hand. With a small and very thin stick he scrapes off a little of the soft glue from the glue stick and runs it along under each feather, pressing the feather down into its new-formed glue bed by running his thumb up the feather toward the nock.

The arrow was now leaned against the log again, near the fire to dry. After about three arrows had been thus glued, the glue stick had to be dipped in the water and held again over the fire as it hardened quickly when cool.

An arrow so glued dried in about half an hour.

A flat board was then picked up—or anything else with a flat surface. The arrow was laid on it and the feathers trimmed, one after another, with a knife. A little projection was left untrimmed at either end of the feather, to correct any slight irregularity in the trimming, which might otherwise prevent a perfect flight.

Arrows had to be taken care of carefully. Dew, rain, or water from a leaky roof might easily warp arrows, or the sinew bindings might get loose. To repair the bindings, the sinews were not usually removed, but merely pressed down with the thumb while the shaft was revolved in the hand and then let dry again.

Warped arrows were straightened by being put in the mouth and bent back true again. In winter or stormy weather they were then dried by the fire, but it was thought better to dry them in the sun if it was possible because fire was apt to spoil the glue.

In my time there were few good arrow makers in the tribe, and the younger men showed little disposition to keep up the art. Arrow makers in old times, we believed, did not have good eyes, as the scrapings from the eagle feathers got in their eyes and injured them.

An arrow maker would make ten arrows and a bow and give it to his wife saying, "Take these and give them to one of your brothers." By brothers he meant either her own brothers or some member of her or his own clan or society. She would do so, and the one receiving the gift would say, "Thank you, sister." And he would give the arrow maker a horse for the gift.

In battle if a man was very brave—because sometimes two brave men went very close to the enemy—much closer than the rest of the party—if either of these men spent all their arrows they could easily get more, as all the rest of the party would be glad to help refill their quivers. They did this because the two men were very, very brave. Only two or three men would be found willing to go so near the enemy.

When the enemy's arrows came toward us in battle, we never picked them up. We thought, "If you do so, then an arrow will hit you, for you have touched one of the enemy's arrows."

But if we were chasing the enemy and shot at them, we did not fear to pick up again our own arrows that we had shot.

In my time, we used iron heads, but when my father was young they still used flint heads. Flint arrows were put into the quiver, feathers down, with the heads sticking up out of the quiver. This was done because the flint points, rattling against one another, were apt to break off. As a protection, a bit of buffalo wool from the buffalo's head was twisted about each arrowhead. When the arrow was withdrawn from the quiver, the tuft of buffalo wool was jerked off.

You will notice that the bird-arrow that I have made for you and now hand to you is feathered with one long feather bound on flat and with one very short one. We thought this enough. The heavy head weighted the shaft sufficiently with the one feather and the small one on the other side to balance it. If we put on two long feathers we thought it was apt to give the arrow a heavy, sluggish flight.

The arrows of our tribe were made rather heavy so that they would fly safely in our heavy prairie wind.

Hairy Coat's father used to shoot a short arrow with a head made of the blade of a butcher knife. Such an arrow would not carry far, but when it hit a buffalo it nearly cut the animal's insides in two.

An arrow should measure about six fist widths, or a little more, not counting the head. With the head it should measure about seven fist widths. Another way to measure an arrow was from the shoulder joint to the tip of the thumb, which every man will find to be just about seven times the width of his own fist held as one grasps a stick in the hand. An arrow that was too long or too short did not have a good flight.

It was my father who told me that before our tribe used flint arrows they used arrows with wooden points. Also that in old times they made arrowheads of *itsutsa*, the big yellow tendon in a buffalo's neck. They also made whole arrows of *itsutsa*. Heads, I have heard, were made of buffalo horn, but I never knew but one man who used horn arrowheads. I have forgotten his name just now. He used arrowheads of buffalo horn because of a vision.

A story says that a boy was taught in a dream by the gods that with buffalo horn arrows he would kill an enemy. Other men, however, did not use buffalo horn heads, just this one man who wanted to kill an enemy.

I never heard of bone arrowheads being made by our tribe. But I was told by my father that we used to make arrowheads of beaver teeth. I never saw such arrowheads myself; however, the beaver teeth were said to have been boiled in a kettle for a long time, were then taken out, and pressed with a heavy stone. The process of boiling and pressing was repeated until the tooth was made straight. A beaver tooth was naturally sharp and did not have to be ground. It would go right into the flesh of an animal. All this I learned from my father, Small Ankle.

(Bone or horn points, apparently arrowheads, have been found in the old Mandan refuse heaps on old village sites near Mandan ND. GLW)

An arrow with a spiral feather was called *imudumite*, or wing twisting around. We did not say "arrow-feather" but "arrow-wing" in speaking of the feathers of an arrow.

We thought arrows perfectly winged when they were feathered with eagles' feathers. These were usually wing feathers. Tail feathers were very expensive, and only a few rich families could afford to use them on arrows. Poor men used feathers of ravens and of other birds to feather their arrows.

In autumn, hunts were organized in the villages to catch eagles. The hunters went to the mouth of the Little Missouri in the badlands or on the big bluffs of the Missouri.

The eagle feathers that were brought home from these hunts were carefully kept in readiness for use.

Spiral feathered arrows, such as I just described above, were the first kind of feathered arrows that a boy shot. We would say to the bow, "This is Adapozis, Burnt Arrow, and the arrow should fly straight. You should keep this arrow sacred and pray to it." We said this because of the story of Burnt Arrow, which I have already told you.

There were a few men in the tribe who always carried two of these spiral feathered arrows in their quivers. These arrows they would not ordinarily use. But when they came closer to the enemy, a man having these spiritual arrows would take them out and pray to them, "Kill this enemy," and he would shoot at the enemy with one of these arrows. In my time I never saw this custom used, but I have heard of it as being in our tribe in former days.

In battle, when we came to close quarters and had to shoot rapidly, two arrows were grasped in the teeth, feathers to the right; a third arrow was grasped in the left hand against the back of the bow slightly athwart the bow with the feathers down and appearing below the hand a little to the left. A fourth arrow lay on the bow string. Sometimes a fifth was grasped in the left hand in addition to the one already held, but instead of being pressed against the bow as was its fellow, it was held between the first and second finger. Held this way, however, it was apt to interfere with the grasp of the bow.

In shooting in battle, the arrow on the bowstring was used

first, then those held in the teeth, lastly the one grasped against the back of the bow.

I have made you some bird arrows side tipped with quills and thorns. As boys, when we hunted birds, we used the arrows with big blunt heads when shooting at anything on the ground and for general shooting. These quill and thorn arrows were more valuable and the thorn and quill points easily broken. Such arrows were used for shooting into a tree, because the arrow would come down and stick upright in the ground and be easily found. A blunt headed arrow would fall flat and not easily be found. Also we used quill and thorn arrows to shoot into flocks of birds, often bringing down two or three birds.

Arrows snake readily if shot at anything sitting on the ground. This snaking was apt to break off the thorn or quill points if this form of arrow were used.

In old times, when one went on a war party, it was a common thing for a warrior to carry two blunt headed arrows. With these he knocked over birds and gophers if he was hungry.

Types of arrows

Wolf Chief (vol. 14, pt. 2, 1913: 444–47)

I had a bow of chokecherry wood—a selfbow, unbacked, and a quiver of arrows. I carried them on my back. The bow case and quiver were in one piece.

[A selfbow is made from a single piece of wood.]

As I started off on my mare, my father said to me "Now if you meet any enemies while you are guarding your horses try to get away and escape home. But if you cannot, stand up against the enemies like a man and try to make good use of your arrows."

I had three kinds of arrows in my quiver. Five or six of them were pointed with iron heads and feathered with prairie chicken feathers like A. Two were blunt headed arrows like C, and seven or eight were wooden arrows with pointed shafts like B. These last two kinds were feathered with either duck or owl feathers.

The blunt headed arrows were for birds. The iron headed

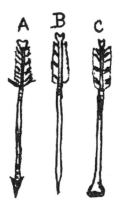

49. The three types of arrows. Drawing by Goodbird. (Courtesy of the Minnesota Historical Society)

arrows were for enemies. The arrows with wooden points were for gophers and small game. We also played arrow games with the pointed wooden arrows—the kind shown by B. These sharp wooden arrows were prepared by shaving the end of the shaft to a point, oiling it with fat and then holding it over a fire of coals to harden it. The other arrows were not treated this way.

The blunt headed arrows were of chokecherry wood. The others were of juneberry shoots. Juneberry wood is dense and heavy. Chokecherry wood is lighter and better suited for the rather clumsy, blunt headed shafts. A dense, heavy wood would have made these arrows too heavy.

We also used another kind of arrow for shooting at birds. It had thorns bound to the side of the shaft near the point. I did not take any of these arrows with me this day because they were not much used but on the prairie. Their chief use was for shooting through flocks of birds or at a bird up in a tree; and besides, to have carried thorn arrows in my quiver would have injured the feathers of the others, for the thorns would have scratched the plumes as I drew them from the quiver.

My father had made all these arrows for me, and they were therefore quite handsome for he was an arrow maker. They were

50. An unstrung bow in its case. The individual carries the quiver of arrows as he would when not in combat. A quirt is in his right hand. Drawing by Goodbird. (Courtesy of the Minnesota Historical Society)

feathered, as I have said, with feathers of an owl wing, a prairie chicken, or a duck.

My father said, "These arrows feathered with prairie chicken feathers are the best. They will fly swiftly." He feathered the five or six iron pointed arrows with prairie chicken feathers—three plumes to a shaft. The other arrows were irregularly feathered. Some of them had all three plumes of owl or duck feathers, others had some duck and some owl feathers.

My bow was thrust into the quiver with the loose loop of the bow string out. The strap from which the quiver was slung was carried over my left shoulder and under my right arm. In battle the quiver would have been turned with the arrows in front. Sometimes in battle the quiver was tied to the belt.

Bows

Wolf Chief explains and demonstrates the craft of bow making to Gilbert Wilson (vol. 10, 1911: 1–3)

We made bows at any time of the year, either summer or winter. I cut a small ash this morning and trimmed it to shape. A good bow staff should be free of knots, and this one, as you see, is nearly so. The bow that I am making I have trimmed down into its final shape, although it needs to be polished. I have bent it into its proper shape and forced it home between pegs driven into the ground to dry. In old times when we made a bow in winter, we lashed it to a long piece of wood. We then inserted pegs to form the curves and to keep it in shape while it dried by the fire. In summer we did as I am doing now.

The bow staff we cut with the back to the bark. But the wood immediately under the bark, being soft, we shaved down to the depth of about three-eighths of an inch. A bow cut in any other way, with too little of the heart of the tree cut away, was apt to crack.

A bow should be about twelve fist-widths in length; that is, it should be about twelve times the width of the fist clenched around the bow. If longer, the bow will be weak, and the arrow will tend to fall short.

Bows were made of young ash, chokecherry wood, or wild plum. Cedar made an excellent bow. It cast an arrow well but was open to objection because it was easily broken. A wood that we Indians call white wood was also used. It is a white, firm wood but was scarce in this region and hard to get. Elm was also used but was likely to split and had a rather weak cast. White wood looks like elm but never grows large—only about the size of a chokecherry tree with a trunk about as thick as one's wrist.

51. Setting the curve in a bow. Drawing by Gilbert Wilson. (Courtesy of the Minnesota Historical Society)

(Note: By white wood, Wolf Chief seems to mean a species of birch which grows, but in scant quantities, in the region of Fort Berthold Reservation. The elm here mentioned I could not verify, but suspect it to be a species of what I would call a large willow. Of this, however, I cannot yet be sure. GLW)

[White wood might be hophornbeam (*Ostrya virginiana*). Stevens (1963: appendix) notes that a few specimens are known from McClean County, which encompasses a large part of Fort Berthold Reservation east of the Missouri River. It is a small tree with very dense wood and looks somewhat like elm.]

[The elm was most probably an American elm [*Ulmus Americana*], which was not uncommon prior to the Dutch elm disease in the 1970s and 1980s. Both birch and willow wood are easily broken.]

Of these woods, ash was much the best. Besides being a good casting wood, we prized it because in war the bow, as a last resort, might often have to be used as a club. An ash bow would stand a very heavy blow without breaking. As the bow would nearly always be braced when occasion arose demanding it be used as a club, the strong bow string helped break the force of the bow. It was really surprising what a heavy blow a good ash bow would stand. This bow that I am making for you is not as heavy as they were often made in old times when their use as clubs was more often kept in mind. But this bow is quite strong enough for either war or buffalo hunting. After ash, we considered chokecherry to be the best wood for bows.

A green bow staff, newly trimmed to shape and put, as you see here, in its drying frame was let partly dry. On a warm day like this, drying required six or eight hours. It was then removed and oiled with a piece of cooked suet, which is more oily than raw suet. The purpose of the oiling is to prevent the bow from cracking as it dries.

Arrows for boys

Buffalobird-woman and Goodbird (vol. 26, 1911: 133–38)

Girls were not taught to shoot the bow and arrow, but boys began to shoot at about three years of age. At first boys were given only grass arrows. There is tall red grass with large joints [big bluestem] that grows on this reservation. The stems of this grass were gathered for boys' arrows and were at their best about the time ears began to form on the corn—the first week of August. The boys made up little parties to go into the hills and gather the grass stalks.

A long leaf growing at the joint was left on the arrow to serve as a feather. A grass arrow lasted only a day or two when it became dry and useless. Grass arrows were shot about within the earthlodge or anywhere.

The bow put in the hands of a very small boy was a chokecherry sapling or ash and about a foot long. Such a bow and grass arrows were used by a boy until he was about seven years old. At the latter age a boy was given a boy's bow about two feet long instead of his small boy's bow.

I made a bow for my son Goodbird when he was a little boy, and I made him a quiver of a prairie dog skin for his grass arrows.

Arrows of buckbrush

In winter, when grass arrows could not be had, very small boys had arrows made from thin shoots of broombrush [buckbrush]. The boys or someone would bring in the new cut shoots and the boy's mother would take them and thrust them one by one into the hot ashes. When the shoot was hot she drew it out and taking a piece of buffalo skin in her left hand she closed it around the shoot and twisted the shoot thereby loosening the bark which now slipped off. We knew the shoot was hot and ready to be

52. A boy with his grass arrows, quiver, and bow. Drawing by Goodbird.
(Courtesy of the American Museum of Natural History)

drawn from the ashes when we heard the bark crackle, c-c! (sh-sh!). All this was done after the shoot was trimmed. Buckbrush arrows were not feathered.

My son Goodbird shot me twice with such arrows when he was small—once in the arm and once upon the cheek.

Forms of arrows

Arrows for boys of seven years of age did not have iron heads. They were usually made blunt with a large round head of the natural wood for shooting birds because these did not break the flesh. Arrows for flight shooting or for shooting gophers were pointed by shaving the end sharp and sometimes hardening it in the fire.

Added to the foregoing by Goodbird

Bird arrows were also mounted with thorns—three of these being bound on the shaft just back of the point. We often shot into a flock of blackbirds as they rose and frequently killed one with either a blunt or a thorn spiked arrow.

Blunt headed arrows were made from chokecherry shoots. Sharp pointed arrows for gophers or rabbits were of juneberry shoots. These last were heavier and carried better in the wind than those of the same design of lighter wood.

Bird arrows mounted with thorns were common. The thorns were from wild plum thorns and had their points cut and somewhat shortened. My grandfather also made some arrows of similar design with bird quills bent double and bound to the shaft behind the point like the thorns.

Arrows of boys seven and up were feathered. A single feather might be split and mounted and glued spirally around the shaft; or the shaft might bear three feathers mounted in the usual way; or two feathers might be bound on the shaft flat, one on either side, without glue; or two feathers split might be put on in the usual way. This last method was common when one had but a scant supply of feathers. It was common for boys' arrows to be feathered without glue, the plumes being bound in place with sinew.

We did not use prairie chicken feathers for arrows. Arrows so feathered we thought hid in the grass, for prairie chickens are adept at hiding in the long grass. We feathered our arrows with eagles' feathers, ducks' and geese feathers and hawks' feathers. Of course we boys could not get eagle feathers. They were brought in by the fall hunting parties.

We boys thought three or four arrows quite enough for one to have for a day's bird hunt.

Our bows were made with the upper arm more springy than the lower arm. We sometimes kept account of the birds we killed by cutting little notches in the upper arm of the bow or on the bracer.

Little boys were not always careful of their bows and used them until the bows got old and bent.

A song of the bow

As boys went out to hunt birds, the little girls of like age often sang a song to tease the boys. Translated the words are as follows:

53. Boys hunting with blunt arrows. Drawing by Goodbird.
(Courtesy of the Minnesota Historical Society)

Those boys are all alike;
Your bow is like a bent basket standard;
Your arrow is fit only to shoot into the air;
Poor boys, you have to go barefoot.

[The standard is the frame of a basket.]

Mock battle with grass arrows

Wolf Chief (vol. 26, 1915: 111–19)

Grass arrows

I think I was about four years old when I owned my first bow, and it was of chokecherry wood. I shot grass arrows which were often carried in little quivers of prairiedog skins. Arrows were of red grass of which are two kinds, one kind growing down in the flats [big bluestem—*Andropogon gerardii*], the other on the sides of hills [probably little bluestem–*Andropogon schizachyrium*] and both good for making arrows. These were about eight inches long and were shot heavier end forward. A grass arrow was so cut that a tiny leaf was left at the lesser end to act as a feather. At six years of age I often played battle with other boys using bows and grass arrows. We commonly divided into two sides, but these had no leaders or officers as you saw white children choose. We began our play very good-naturedly, but as the two sides drew close, some were pretty sure to get hurt. And the mimic battle often ended in a quarrel—though Indian children never fought with their fists as you say white children do. A well shot grass arrow on the bare flesh hurt and often caused the flesh to swell up a trifle. I have wept from the smart of such a wound more than once. Some of my friends are still living with whom I fought with grass arrows when we were boys.

A mock battle

I recollect such a battle late one autumn. I do not remember all the details clearly, but they were about as follows:

About ten of us, boys from five to seven or eight years of age, were playing in the afternoon on the prairie outside our village.

Two boys, Two Bulls and Stands Up, were a little older than the rest, perhaps eight years old. Two Bulls was a brave boy, and he grew up to be a brave warrior, killing three enemies but was himself killed in battle when about thirty-four years old.

Our little company went down to a flat where there was much red grass, and each gathered grass stems for about thirty arrows so that we might have a prolonged fight. A grass arrow made no lasting wound but did make a little red mark on the skin like a flea bite. I had a cloth quiver which I filled and slung from my shoulders over my breast. We divided into two sides and stood about eight yards apart and shot.

Sometimes several of us boys would single out one on the other side, give chase, and catch up with him and shoot him with our arrows. A little five-year-old boy on my side was shot in the ear and wept crying out, "Oh, mother." The rest of us then quit and went home.

The quarrel of Two Bulls and Stands Up

Two Bulls and Stands Up were once on opposite sides and were shooting grass arrows at each other. They shot, coming closer and closer to one another until both lost their tempers and began hitting each other with their bows. At last they dropped their bows and seized each other by the hair, tears running down their faces as they cried, "*Ixka et duca*" or "Let me go." The rest of us crowded around the two boys, crying, "Don't do that. That is bad." But they struggled back and forth until one loosed himself from the other, and both ran home weeping to their mothers. The rest of us also went home.

How the Hidatsa boys fought

Indian boys never fought with fists nor by kicking but always by pulling hair. If a boy got into a fight, his mother would say, "You foolish boy, it was your own fault. You yourself were fighting. Why do you weep?" But our fathers used to say to us, "If you get into a fight with your friend, you must not weep. Fight back, but be a man and do not cry."

Sometimes we threw stones at one another, but the men in the village always stopped this. "That is no way for you to play," they would say. "You might hurt a friend's eye with a stone."

When little boys were fighting with grass arrows, the men, even when watching them, never interfered. They just laughed and called out, "Don't run away, don't run away, but don't shoot at the other's face, shoot at his body. Those grass arrows are sharp and you might injure him if you shoot him in his face."

SIXTEEN

Earthlodges

Building an earthlodge

In reading about the earthlodge, one should keep in mind that the lodge was more than shelter. It was both microcosm and macrocosm of the Earth and the universe as perceived by the Hidatsas. The interior of the lodge was always oriented, in the minds of the occupants, to the cardinal directions no matter what the actual orientation of the lodge.

If one looks at a plan, it can be seen that the earthlodge is, in reality, square. But it is a square house constructed so that the sides are flared out, thereby increasing the floor space. This architectural modification was probably made as various Northern Plains groups evolved into extended matrilocal or patrilocal social units. As many as twenty people lived in the lodges built by the Hidatsas.

After Wilson's death in 1930, the Division of Anthropology of the American Museum of Natural History asked one of its staff members, Bella Weitzner, to compile Gilbert and Frederick Wilson's field notes on the construction and maintenance of earthlodges as a monograph. This was published in 1934 as *The Hidatsa Earthlodge*.

Wilson was fortunate in many ways to have gotten to know Buffalobird-woman because, as it turned out, she was one of the few women in the village to have the right and the honor to oversee the construction of the core framing of an earthlodge.

Unfortunately, Weitzner did what Wilson always sought to avoid: she turned Buffalobird-woman's vivid, firsthand descriptions into a narrative in the passive voice. To return Buffalobird-woman's voice of pride and authority, I am including here some substantial portions of her narrative as transcribed by Wilson.

Buffalobird-woman was very proud both of her family's status and her own. She was justifiably proud of her right to oversee the framing of a lodge, a very exacting task. She gives Wilson this account as an authority, and it is obvious that she wanted Wilson to convey this. By 1909 they had been working together for three years.

So much of the Hidatsa world was constructed of wood, and it was so important to the Hidatsa way of life that at one point Buffalobird-woman says, "We honored wood, we Hidatsa people."

Buffalobird-woman (vol. 26, 1909: 242–43)

Buffalobird-woman's right and how obtained

My right to prepare the four great posts and beams I got from my mother. I gave her one whole suit of clothes of finely tanned skins for the right. Everybody in the village recognized my right, but it was thought not the right thing for me to refuse anyone who wanted to hire my services.

For my work I was paid one newly tanned buffalo skin carefully dressed and soft. The owner of the earthlodge also gave me a big dinner, dried meat, and other things. A big wooden bowl full.

Only a very, very few women in our village owned the right to trim and oversee the raising of the four great posts of an earthlodge.

Buffalobird-woman (vol. 8, 1909: 49–67)

When a man wanted to build an earthlodge, he first got together a great deal of meat and other things to eat and much bone-grease. This bone-grease or marrow was gotten from the inside of the bones and was obtained by pounding up the bones and boiling and skimming off the grease with a horn spoon. The bones of one buffalo treated this way would yield about five pounds or two quarts of this kind of edible fat. Sometimes the man collected the fat out of the bones of several buffaloes before he thought he had enough.

When the man was ready to begin building the frame of the house, all his friends came together into a bee to help him. The meat and other things the man had collected were to feast these friends. But before the frame of the house could be raised, the four great supporting posts upon which the house was stayed had to be raised into place. Only certain ones in the village had the right to put up these four supporting posts and the four logs or stringers that lay on the posts. This was because we thought the four posts were sacred.

I was one of those who had the right to do this—raise the four posts and the four logs on the posts.

This is how I did it.

I leveled off a squared place on the ground and at the corners marked the places for the four posts. We dug holes for the posts with digging-sticks. These were just like the digging-sticks that we used in cultivating our gardens except that they were longer. We drew out the dirt from the holes with our hands, and when the holes got too deep for us to reach easily to the bottom, we drew the dirt out with large spoons of buffalo or mountain sheep horn.

The holes were dug with care, just large enough for the posts to slip into. Each post was carefully measured for this purpose and its hole dug exactly to meet the measure. This was not only to secure firmness for the post by seeing that the post fit tightly in the hole but also in order that the tops of the posts, when the four were all raised in place, stood all at exactly the same height from the ground.

As I have said, the posts were four and were very heavy, and there were four logs to go on the posts. (These logs will in this paper be called the interior stringers. GLW) These posts and logs, or interior stringers, had all been cut the summer before and let lie to cure. They were brought to the village in the winter when snow lay on the ground. This work was done by men. They dragged the posts and heavy logs over the snow with rawhide ropes.

It was my particular work to trim and make ready the four posts and the four logs or interior stringers and to oversee their being raised into place.

With an ax I first cut a fork, or notch, in the top of each of the four posts. The notch was shallow and the two short projections which made the fork were called the ears. These ears were only about three inches long, and this of course was also the depth of the notch or cut that made the fork. In these forks or notches the two lower of the four interior stringers were to rest. We commonly cut the notch or fork in the root or butt end of the post because the trunk was wider at this end and gave a greater surface on which to work.

I next made ready the ends of the four interior stringers. Two of these were to be upper stringers and two were to be lower stringers. The lower stringers were heavier than the upper ones. The lower stringers rested in the forks of the big supporting posts, held in place by the ears which made the fork. The ends of these two logs or lower stringers were cut square for a little way back and each end made to fit into its fork snugly. I worked carefully, measuring with a little stick in order to be exact.

The two upper stringers were not cut like the lower ones but terminated at each end in a rounded nose to let the roof poles [rafters] lie in place. Also on the underside at each end was cut out a place the width of the stringer upon which it rested and leaving a heel which rested against the lower stringer. This prevented slipping.

As I squared the ends of the lower stringers or trimmed the ends of the upper stringers I had to be careful that the flat surfaces I was making were true. To test them I took a little round stick and rolled it back and forth over the surface. By this I could tell if any little inequalities had been left on the surface. It was important that the work be true. Indeed all my work on posts and stringers alike was made with careful regard to measurements.

The surfaces of the posts and stringers were smoothed off neatly; any little projections were cut off, and the bark peeled off. A part of this work, as peeling of the bark, would be done before the posts were brought into the village.

54. Buffalobird-woman measuring the placement of one of the vertical posts (at her right toe) around the periphery of the lodge. Drawing by Goodbird. (Courtesy of the American Museum of Natural History)

All this trimming of the posts and stringers I did with an ax. The little chips that I made in cutting were not touched with the hands but were removed with a buffalo bone. This was because we were taught that logs have worms, and we feared the worms might enter our bodies.

It was not my part to raise the posts and stringers with my own hands. When I had the posts and stringers all trimmed and ready and the post holes dug, the man who was building the house called a number of young men and said to them, "I want to ask you to put the posts in the holes and raise the stringers on the posts."

The man (for whom the lodge was being built) had food all cooked and prepared for the young men, for that is the way he paid them.

This is how the posts were raised. A number of poles were brought like tent poles, only a little heavier. These were laid down in pairs and joined together at one end by a short rawhide rope or cable, thus making a kind of sling. A number of these slings having been prepared, they were taken up by the young men, two young men to each sling, or one young man to each pole. They raised the posts, one by one, and slid them into the holes by means of these

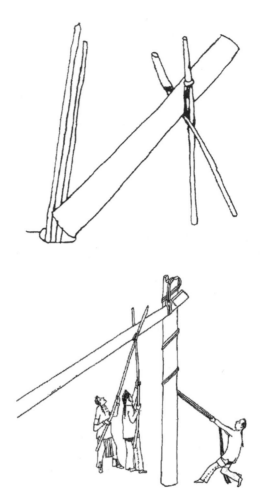

55. (*top*) Placement of one of the four vertical central posts of a lodge. Drawing by Goodbird. (Courtesy of the American Museum of Natural History)

56. (*bottom*) Raising a horizontal stringer onto one of the four central posts. Drawing by Goodbird. (Courtesy of the American Museum of Natural History)

slings. In hoisting the posts the two poles of the slings were held crossed. Besides the young men who managed the slings, others lifted at the posts with their hands and some held the lower end of the post down to the hole into which it was to fall.

As each post fell into its hole, it was twisted around so that it

57. Two horizontal stringers in place on a vertical post. Adapted from a drawing by Frederick N. Wilson. (Courtesy of the American Museum of Natural History)

stood just right to receive the lower stringers [horizontal beams]. As these stringers were to the right and left of the fireplace as one entered the door of the lodge, the posts must stand so that their ears would properly receive these lower stringers in this position.

The four posts having been placed, the two lower of the four interior stringers were raised into position. A rawhide rope was thrown over the fork of one of the standing posts and looped over the end of one of the stringers. Some of the young men pulled at the rope while others manned the slings. In this way both lower stringers were hoisted into place.

The upper stringers were raised in the same way except that the rope could not be drawn through the forks of the posts as the lower stringers now lay in these forks.

The four interior stringers were so placed that when the lodge was completed the two uppermost lay the one on the side toward the door and the other on the side toward the rear part of the lodge. The two lower stringers were on the other two sides to the right and left of the fireplace as one entered the lodge.

Although the posts and the interior stringers were now raised and in place, the post holes were yet left unfilled. This was done

58. The framing of a 13-post lodge. There are 13 outer posts with stringers running from one to the next; 4 central posts and 4 stringers; and 84 rafters. The entryway also requires framing as well. Note that the drawing does not show framing on the sloping sides—the *atuti*. Adapted from a drawing by Frederick N. Wilson. (Courtesy of the American Museum of Natural History)

so that after the stringers were raised, the posts could be moved about slightly and adjusted to any little error of measurement that may have been made. Earth was then shoveled into the post holes and packed firm and solid.

The exterior poles were twelve. The two posts at the door and the two posts to the right and left of them (that is, four posts in all) and also the posts that stood in the rear part of the lodge were a little higher than the other exterior posts of the lodge. In a lodge of the ordinary size the door posts and its fellows of equal height must be of a height equal to that which a woman is able to reach with her arm and hand extended straight up above her head. The shorter exterior posts were less by the length of a woman's palm, measuring from the wrist to the tip of the middle finger.

59. This side view shows the framing of the steeply sloping *atuti* as well as the placement of the posts and stringers supporting the rafters around the lodge. Adapted from a drawing by Frederick N. Wilson. (Courtesy of the American Museum of Natural History)

On another occasion Buffalobird-woman describes the size of these posts.

Buffalobird-woman (vol. 22, 1918: 29)

The four central posts

My mother, Red Blossom [a sister of her biological mother, Wants To Be A Woman, but known as a mother because she was also married to Buffalobird-woman's father, Small Ankle], measured the length of the four central posts with her arms. Each post as it lay on the ground should be two and a half Indian fathoms long. A fathom by my arm is a little more than five feet [about 1.5 m], but Red Blossom's arms were longer, and a fathom by her was probably five and a half feet, so I think the posts were about 13.75 feet [a little more than 4 m]. Of this length, the post was set into the ground to a depth as far as I could reach with my arm and use a bighorn spoon to take out the earth—a depth, that is, of about 27 inches [almost 70 cm]. So the post stood about 11 feet 6 inches high above the ground [about 3.5 m]. For the four central posts, we used big posts—as much as 18 inches [about 45 cm] in diameter.

[A log for the central supports might be almost 14 feet (about 4.2 m) long and a foot (30 cm) or more in diameter. Such a log would weigh more than 300 pounds (about 135 k) and perhaps as much as 600 pounds (about 270 k) if it were 18 inches (45 cm) in diameter. Cottonwood has a density of roughly 25–30 pounds per cubic foot (about 90 k/m³).]

After the posts were in place it took about six days more to complete the house. As I have said, excepting the first part—the raising of the four posts and four stringers—the work was carried on as a bee even when the actual raising was done by the young men.

Ceremonies and beliefs concerning the posts

We honored wood, we Hidatsa people, and we said, "The four great posts have life." And we prayed to them and made offerings to them. In old times we bound neatly about each of the four great supporting posts a buffalo skin or buffalo-calf skin carefully dressed and trimmed square so that it would look neat on the post, or else we bound other offerings on the posts. In later times we used bright colored calico and other cloths for offerings instead of buffalo skins.

We Hidatsas thought an earthlodge was something sacred, and we honored the four great posts because the lodge was supported by them.

The offerings made to the four posts were bound in place and a prayer said after the earthlodge was all finished and completed.

Such were the houses we built in old times. When we lived in them our people were healthy, and there were many children and many old people in the village. But now there are only a few old people left, and the children die. In those days we remembered happenings as having been "in the year that such-and-such an old person died." But now the old people die so fast that we no longer try to mark the years by their deaths.

We had good times then. Always drums were beating, and we feasted. I am an old woman, almost seventy years old. I think of those good times, but I know I shall never see them again.

[Hairy Coat's lodge was the last earthlodge to be lived in. All lodges were treated with a great deal of respect, and in this case much negotiating was necessary to obtain access. This lodge is described in Wilson's monograph, *The Hidatsa Earth Lodge* (1934).]

Frederick N. Wilson (vol. 12, 1912: 2)

Hairy Coat's offering to his lodge

When I was admitted to the lodge Hairy Coat took down from a shelf on one of the main posts a broken piece of pottery, put sweetgrass in it, and lighted this with a coal from the fire. He then placed it before his shrines and prayed to them, asking them not to be offended at this whiteman coming into their lodge, that he meant no harm and was trying to learn about Indian ways so that these ways would not be forgotten. After this prayer he announced that it would be all right, and I then proceeded to photograph and measure the lodge. The prayer was rather impressive as Hairy Coat talked to the shrine with gestures, just as he would to a man, and, in fact, acted just as though he were speaking to a living "presence."

On Earthlodges (The observations
of Hairy Coat and Not A Woman)

Hairy Coat (vol. 12-31, 1912: frame 0699)

Hairy Coat's father's lodge when he first came to Ft. Berthold

My father, when he came to old Ft. Berthold [Like-a-Fishhook Village], built his lodge to be just a little larger than the one I now occupy and then his second lodge to be a little larger than that. He came to old Ft. Berthold when I was eight years old—after leaving the Five Villages [the three Hidatsa and two Mandan villages near the mouth of the Knife River]. We spent the winter at the winter village. I was seven years old when we left the Five Villages—this was 1844 [1845]. We came in the summer time, and the first winter all our horses were stolen by enemies.

This lodge had a flat roof and had sixteen posts around the *atuti*, but the logs were larger than in my lodge here because the lodge was larger. Other people often made their lodges of sixteen posts—these were the largest and best. Other smaller lodges had twelve and fourteen posts.

Hairy Coat (vol. 12-31, 1912: frame 0708)

It was a woman's place to build lodges, and in my father's first lodge at Ft. Berthold my mother cut the timbers. The man would say what size lodge he wanted, and then the woman would go out and cut them. The logs found on Missouri River sandbars all last longer—the cut timber has worms of itself—but the river timber lasts longer although it has them, too.

Hairy Coat and Not A Woman (vol. 12, 1912: 26–29):

Dismantling and rebuilding an earthlodge

As a rule a lodge lasted seven years and was then rebuilt.

In taking down an earthlodge, so as to rebuild, we began at the smoke hole and with our hoes pulled the earth down to the ground on all sides, working outwards from the smoke hole. After the earth was pulled down, we took off the dry grass and then the willows and boards at the smoke hole, then the cross stringers over the fireplace (at the smoke hole). After this was done, we removed the dry grass and the willows on the rafters and took down the rafters or lodge-ribs. And lastly we removed the dirt, the willows, and the poles at the *atuti*. To remove the main stringers, which were sometimes used again, we just took a pole and pushed them up and allowed them to drop. Should they remain good after the fall, we used them again.

If the main posts were old, we just pushed them down because after seven years they were usually rotten at the base and just broke off. But if the post was solid and did not break off, we dug at one side and worked it about until it was loose enough to take out. We did it this way to keep the post holes as small as possible. We wanted the post holes so that the posts would just fit.

The outer or *atuti* posts [the outer ring of vertical posts], being smaller, always broke easily and needed only a push to go over. The logs or posts were usually all rotten at the end of seven years. The rafters or *atidutidu* poles [the four center poles] were all saved to make *atuti* poles—mixed with new poles if necessary. We always renewed the rafters entirely.

All the other wood, posts, etc., with the exception sometimes of the four stringers, was used for firewood.

After the main posts and *atuti* posts were up, and the stringers placed, the *atuti* poles were placed in position. These poles were set with the large end down, smaller up.

If there were any helpers working on the lodge, they were entitled to all cut off ends and all trimmings left in preparing the *atuti* or *atidutidu* poles.

The poles for the roof (the rafters) were selected and divided into four piles. The best poles were then placed first over the door, as it was here that anyone went up onto the lodge. The rest of the poles were then placed on the other three sides. The small poles were left for the corners because no one walked there— only on the middle parts. In putting up the rafters, usually two women did the work.

An *atidutidu* pole was placed with upper end on the central stringer first, and then one or both lifted the lower end and set it in place on the *atuti* stringer. One of the women then, with a forked stick or pole, adjusted the poles evenly at the top.

If the ground had been worn down a good deal, the outer or *atuti* posts were cut a little longer than in the first building.

The cross stringers (on the central four posts) were measured at the distance between the bottoms of the main posts. Then a log was cut and used as the measure. The largest and best logs were cut as cross stringers for the part close to the smoke hole and the smaller ones for the parts farther away. [They are describing here the construction of a flat-topped lodge, which requires more intricate cutting for framing. The traditional lodges had more or less flatly conical roofs. My opinion is that the flat-topped lodges were not built until iron tools were available to do the more intricate woodworking.]

After the *atuti* and *atidutidu* poles are in place, the whole lodge is finished [sheathed] with the willows. Then two men or women go to the roof and, with rawhide ropes, hoist the cross stringers to the smoke hole and adjust them to place. [He is describing the construction of the flat platform on this style of lodge.]

When this is done, they are ready for the boards and willows, then the dry grass and earth for the whole lodge.

The boards (of split logs or slabs) [also called puncheons] were carried to the roof and laid lengthwise of the lodge across the cross stringers and were fitted as close as possible, with ends butting into each other, the longer boards at the smoke hole. Willows were laid lengthwise on these boards to fill up the cracks and then covered with dry grass. More hay was used at the flat part or smoke hole than on any other part of the lodge. This was to keep the rain from leaking through. This hay was the long river grass [cord grass–*Spartina pectinata*] that is tough and lasts a long time.

If split logs were used as boards, the round part was placed uppermost, but often these slabs were flat on both sides.

After the dirt was placed over the fireplace roof, the smoke hole box was then fitted, like the logs at the corners of a log house. The earth was pulled up close to this box edge and raised somewhat so that it would keep the water away from the smoke hole and thus keep it from coming down into the fireplace.

Winter lodges and twin lodges

Villages were constructed on the higher terraces above the river—
the Knife River in the case of the early Hidatsa villages and the
Missouri River in the case of Like-a-Fishhook. This location had
the advantages of being more defensible (the riverbank forming part of the defensive system), more open and expansive, and
more social because everyone was living in the same area. But
during winter the summer villages were exposed to the excru-
ciating, cold winds characteristic of Great Plains winters, and
the larger (about 40–50 feet or 12–15 meters in diameter) summer lodges could not be kept comfortable. Water was a problem
because the Missouri River froze solid, and therefore a source
of running water was a necessity. And it was necessary to have
a ready supply of wood because foraging for wood in the winter
was not always feasible. Therefore, the larger village split into
smaller winter camps that were frequently located in the coulees
(as the large ravines draining onto the Missouri River are called)
or other well-wooded locations not too far from the site of the
summer village because people had to return there to get food
from their cache pits (the grass-lined storage pits).

Winter villages were occupied for just one winter. Lodges were
about 25 feet (about 7.5 m) in diameter, and the construction was
fairly crude (stringers laid in forked posts instead of being carefully fitted). Twin lodges (described below) and hunting lodges
were sometimes built according to essentially the same plan.

Buffalobird-woman (vol. 20, 1916: 149–51)

When I was a girl, when we went into winter camp, many families
would build small earthlodges just like our summer earthlodges.

That one you saw at Owl Woman's was what we called a twin

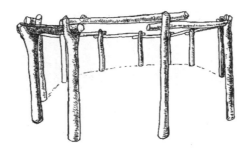

60. Winter lodge framing involved much less precision cutting and trimming than was required for the large earthlodge, which was occupied most of the year. Drawing by Frederick N. Wilson. (Courtesy of the American Museum of Natural History)

lodge. It was built adjoining an earthlodge to which it was connected by an entrance. It was not used by a young man and his wife, but if a young man had aged parents, he built it for them because they needed a warm place. Also a man who had a daughter whom he loved very much built one for her to play in with her girl companions. But no young man and his wife ever lived in one. Twin lodges were very common where there were old people and daughters.

We used twin lodges in our family, and there was a little twin lodge like the one at Owl Woman's that my grandparents lived in. I remember we all used to run in it on a very cold day to keep warm, returning in the evening to our earthlodge. The diagram shows our winter lodge which we had for a short period in the Bent-Enemy-Killed winter [1856–57 (Howard 1960: 35)]—for about three months, I think. We never had a twin lodge in the summer village.

The twin lodge caught fire one year, but we put it out before it burned down.

To make a twin lodge like the one at Owl Woman's to live in for the winter was hardly ever done. Everyone would have made fun of it. But it was very commonly made for living in for a short time on a hunt. It was built by putting up four central beams

61. Structure of a "twin-lodge." Drawing by Frederick N. Wilson. (Courtesy of the American Museum of Natural History)

and posts as for an earthlodge (only the posts and beams were lighter) with long poles lying against them. The roof was covered with willows, then grass, and then earth. The willows went upright and then bark was laid over them (if found in quantity). Then grass and earth were laid on top of them. On the outside, forked posts and railings surrounded the lodge as in the case of the earthlodge. Sometimes for security, there were three or four tiers of these railings and posts. [She is referring to the posts supporting the earth covering to keep it from sliding off.]

Puncheons were split and the slabs made to fit into all spaces above the four beams so that grass could be used up to the smoke hole. The smoke hole was about 20 inches across. It could not be made too small, for light was needed. The entrance to the earthlodge was closed by the hide of a bull or cow buffalo.

A hunting lodge was the size of a large teepee, or about six paces across the floor. (Owl Woman's is 21 feet across the floor. GLW) The four center posts were about 5.5 feet high. A twin lodge was smaller, especially those used as play houses for the girls. The peaked hunting lodges like those used in eagle hunting were also used in ordinary hunts as for deer.

I have built such a lodge as this on two occasions. One I built in the usual way with earth as Goodbird has described. The other time Wolf Chief and White Bull went with my husband and me. White Bull was a Tsitsika [Prairie Chicken clan] and my clan brother. We found the ground so hard that we could not get enough earth to cover the lodge properly. It was the second winter month or December. We made a fire and thawed the

62. A similar use of posts can be seen on this hunting lodge built for shelter by a group catching eagles. Drawing by Goodbird. (Courtesy of the Minnesota Historical Society)

ground and heaped earth around the bottom as best we could and then drew the skin cover of our tipi around the upper part of the lodge. We put posts and rails around it in the usual way.

Once when we were camped for the winter just below Elbo-woods, we found a village site. In one place there had evidently been a very large lodge. Some of us dug down into the earth and found posts of cedar burned at the top but sound at the bottom. We also found a place where the door had been, and Son Of A Star found a fine big clay pot buried in the earth. He dug it out very carefully, but it fell to pieces. We counted six rings or lodge circles.

I have heard that Itsikamahidic taught us how to make the peaked hunting lodges (like those for eagle hunters). Afterwards my people found timber floating down the river. This they gathered, selecting those that were forked on the end which they burned with fire to trim into forked posts. And so they began to build earthlodges. Sometimes also a twin lodge was made like an eagle-hunting lodge. That is, it did not have four central posts and beams upon which the long poles of the roof leaned, but was peaked like a tipi.

The peaked or tipi-shaped hunting lodge

Hunting was often a social event as well as a subsistence strategy. Those hunts described by Buffalobird-woman and Wolf Chief usually involved several families and their infants. Goodbird was born on a hunting trip up the Missouri River. Buffalobird-woman had Goodbird make several drawings showing who slept where in their hunting quarters on particular hunts and still other drawings to illustrate the positions of bullboats and their occupants as the hunting party returned to camp (Wilson 1924).

The skeletons of a few of these hunting lodges are still tucked away in the junipers in the badlands. What is notable is the effort put into what is a temporary shelter. Whether similar lodges were built before iron tools were reasonably abundant is unknown.

Buffalobird-woman (vol. 22, 1918: 23–26)

A peaked hunting lodge was built on a frame of four forked poles, like a Hidatsa tipi. Around this frame, in a circle we placed other poles, setting these as closely together as possible. Many of these poles were long and met together at the top, but some were shorter and placed so as to help close the spaces between the longer poles, especially down near the ground. Of the longer poles, especially if any had branches when first cut, we trimmed the branches so as to leave the stumps standing on the pole. These projecting stumps could be used to uphold horizontally laid sticks which support the earth that was laid on the roof as a final cover.

Two of these stumps of branches, larger than the others, were so placed that they projected into the inside of the lodge. From these a thong swung across the fireplace for drying meat and the like.

The frame of poles being now completed, I began at the bottom and laid a row of buckbrush, chokecherry, elm, red willow, or other green-cut branches quite around the lodge, covering all the spaces between the poles of the lodge frame. For such use I chose only the straighter branches, standing them upright with stem or root end down.

Again beginning at the bottom, I laid over the buckbrush and branches, for a thickness of about 4 inches, the long kind of grass which grows by ponds and streams [cord grass]. I stood this grass against the buckbrush and branches, root end down, working very carefully. This grass was to keep the lodge warm and to prevent the covering earth from falling through into the lodge. Over this grass I heaped earth.

This first row, or series, of buckbrush and branches and grass carried the earth about 4 or 4.5 feet up the sides of the lodge.

The process was now repeated in a second series, until branches and grass and earth were carried around the lodge to a height of about 8 feet. All was earthed over to the depth of about 4 inches. To do this I would take a wooden bowl of earth in my hand, pour the earth on the lodge roof at the proper place and press it down with my palm. Above the height of about 8 feet no earth was laid.

After the earth was all put on, the lodge was encircled by rails, laid in forks of short posts. There were three series of these rails. The use of these other rails was to uphold the earth.

For the space of about 2.5 feet above the earthed over portion of the lodge poles, bark was laid or, if bark could not be obtained, short puncheons were split and used. Around the lower edge the bark was thrust about 4 inches into the earth covering of the lodge in order that the bark might be held firmly.

We used pieces of elm bark chiefly, of which an abundance was found in the badlands where we usually made our hunting lodges. No doubt, down in the Missouri River timber, cotton-

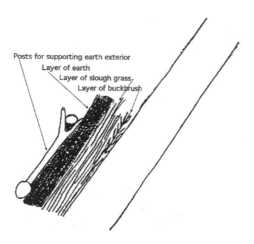

63. Details of the construction of a hunting lodge roof. (Courtesy of the American Museum of Natural History)

wood bark has been used for the same purpose, but it happens that I never helped to build a hunting lodge where plenty of elm bark was not to be found.

The pieces of bark were laid on like shingles, but in reverse order, or else they would have fallen. The bark was laid as it came from the tree, with the rough side to the weather.

I used no ladder to ascend up on the bark, but I stood on the rails that were laid in the forked posts to keep the covering earth from falling down.

Above the bark for about 2 feet to the joint or place where the poles crossed, there was no covering put over the lodge poles. Thus a place was left to let in light and let out smoke. However, a hunting lodge did not get very smoky. The poles through which the smoke passed to escape broke the force of any descending draft that might blow in.

A fair-sized hunting lodge had a floor diameter of about 12 feet. The poles averaged about 3.5 inches in diameter at the greater end, and the height above the fireplace to the joint or crossing of the lodge poles was about 10 feet. These measurements I reckon as the usual size.

When a hunting lodge was made like Owl Woman's, supported on beams and four posts, the size was about the same as above. And a twin lodge, also, was about the same size.

In the lodge that Owl Woman had made, she has put some small slabs of bark around the smoke hole in a horizontal position. This I think is her own idea. We never used to do thus in old times so far as I remember.

Fireplace

The fireplace in a hunting lodge was about 3 feet across and about 4 inches deep. We carried out the ashes quite often. In those days we had few iron tools. Every morning the women carried out the ashes before the fire was made. As we had no shovels, we used a piece of an old rotten log from which the rotten core had been scooped out, leaving the shell. This was used as a kind of scoop, the ashes being raked into it with a hoe. Such a scoop was about 2 feet long and about 6 inches wide and somewhat hollowed inside.

We always had several of these hollowed planks or log pieces in a hunting camp to lay fresh meat on, and several could be laid together, side by side, with the convex side turned uppermost.

We dried much meat in the hunting lodge.

The use of sod as an earthlodge covering

By the time Wilson had returned home to St. Paul, he had thought of some questions about covering the lodges that he had failed to ask. He enlisted a couple of his friends on the reservation to provide him with the missing details. From these remarks, particularly those obtained by Mr. Shultis from Buffalobird-woman, it should be clear that the use of sod to cover a lodge was a recent practice made possible by the use of iron tools, including a plow.

Rev. C. L. Hall (vol. 8, 1909: 68)

When the Indians used to build sod houses, they clubbed together into a bee to raise the building. After the poles and logs were cut, a single day sufficed to raise the frame. I am told that at old Fort Berthold the men used to bring in some of the timbers that were to make the supporting posts, after they had been cut, by hitching a lariat to one end and dragging them by hand one at a time to the village. Horses were scarce in those days.

Sod for the roof was carried thither by women who bore it in baskets. These were the usual bark-woven carrying baskets of this reservation, held by a thong over forehead or shoulders.

Buffalobird-woman—obtained at Wilson's request by Mr. C. A. Shultis (vol. 8, 1909: 69)

In regard to your inquiry concerning the placing of sod in the earthlodges, she says that the grass side is placed down. She remembers years ago in the old villages that no sod was used (it probably being too difficult to get as they had only sticks and bones to work with), but soft earth was carried in baskets by women from some place along the bluff (over the Missouri) or where it was easy to obtain and passed up to other women who emptied them on the roof. If they made the dirt covering thick, the rain seldom penetrated it. When sod was used, enough loose dirt was thrown in to fill the cracks or spaces between the pieces of sod. They use sod more since the advent of the plow, and the men assist in the construction. It was always the women's work in old times and was considered disgraceful for men to engage in it, although sometimes for a lark a party of young fellows (boys in their teens probably) would come around and work for a little while and then skip—just making fun, I presume.

Dismantling an old earthlodge

Because wood, especially wood large enough to use for earthlodge timbers, was scarce and hence valuable, as much wood as possible was salvaged from a lodge in need of reconstruction. Some timbers were always in better shape than others, and the largest and most valuable were those first to decay where they came into contact with the soil. Therefore dismantling a lodge was done with care to conserve as much wood as possible and with almost as much ritual as was part of its original construction. [Some of this narrative has appeared previously, in slightly different form.]

Buffalobird-woman (vol. 22, 1918: 43–49)

Duration of an earthlodge

An earthlodge should last about ten years. If the posts, especially if the big central posts were of big thick trunks and the beams were of good size, the lodge might last twelve years. A lodge whose posts and beams were not very heavy lasted eight or nine years.

One or more of the lodge posts was sure to rot sooner than the others. A post always rotted away at its foot, or the part in the ground, and owing to the weight of the building upon it, the post would sink straight down.

This sinking was a sign that rot had set in. It was of no use to try to replace the sinking post. The other posts were sure to have rotted likewise and would soon sink also.

The decayed and sinking post might be one of the central posts or one of the peripheral posts.

Dismantling the lodge

When it was evident that a lodge must be torn down, the owners, husband and wife, invited five or six of their women relatives (blood) to come and help. The women of the lodge attended to inviting them.

I do not know why, but we always observed that when an earthlodge was torn down, a big wind arose. An earthlodge was akin to a sacred object and seemed to have the power to cause a wind to arise.

For this reason, the women always chose a calm, clear day to tear down a lodge. They came early after breakfast, but before the lodge was all down, a wind was sure to arise.

Time and method of wrecking

The time to tear down a lodge was in the spring, before we had started planting our gardens, which were planted early. The ground often was not yet thawed.

The women brought their hoes and, beginning at the smoke hole, they scraped off the earth from the flat-planked part of the roof. They did this carefully, not scraping the underlying grass layer away. All the earth was thus scraped off the lodge to the ground just outside the lean-to or *atuti* puncheons (forming the lowest part of the lodge).

They bore the grass off in their arms and cast it outside of the village, since it was now useful for nothing.

The willows—for the flat-planked roof was covered with them also—were borne down and divided among the women to be used for firewood. The lodge owner kept some for her fire also.

If it was a flat-roofed lodge, the planks covering the flat roof were also borne down by armfuls and divided among the women.

[As previously noted, flat-roofed lodges were probably not constructed until historic times when iron tools made the extra work of cutting both the additional wood and pieces of sod feasible. The flat area was above the area formed by the four main supporting posts and made a sort of "porch" on top of the lodge. People liked to sit on their lodges just as other people like to sit on their porches or roofs, and making the area flat made it all the easier to sit or stand there.]

The rafters that upheld the planks of the flat roof were rolled down the sloping roof behind the lodge. They were left for the owner, who might use some of them in building her new lodge.

The main rafters, or *atidutidu*, were next taken down. A woman stood outside of the lodge and with a pole about the size of the pole of a dog travois frame she pushed against the end of each rafter as it rested on its exterior beam, so that it fell inward on the floor of the lodge.

All the main rafters were left to the owner. Lying under the roof they had been kept dry, and many of them were sound and could be used on the new lodge.

The lean-tos or *atuti* puncheons were divided up among the women and the owner for firewood.

The four big beams [stringers] on the central posts were dismantled thusly: first the two shorter, upper-lying ones were dislodged by pushing one end with a pole, followed by the two larger, underlying beams, which were set in forks or pairs of "ears" cut in the tops of the vertical posts. To be dislodged, the beam had to be lifted out of one of these forks. Two men did this with a sling made by tying a thong to the tops of two crossed poles. The men were careful to spring aside when the beam fell because, if the beam were lifted off one post only, the jar of its fall often broke off the other post on which it lay and brought it down.

The four big beams on the central posts were preferably made of driftwood trunks salvaged from the Missouri in the June rise. These lasted longer than cottonwood posts, which insects had attacked. Beams made of driftwood trunks might still be sound when taken down and could be used in the new lodge for posts [around the exterior].

Beams (the stringers) on the exterior posts were taken down by two women. They stood on an overturned bullboat and readily lifted a beam down. Sometimes at the raising of the posts of a new lodge two men stood on an overturned bullboat, for such a boat had a strong frame.

The four big central posts were often sound except the end that was in the ground. This rotten end and the ears at the top of the post were cut off, and the posts thus trimmed were used to make the four rearmost exterior posts of the new lodge. Or the four big beams (the central stringers) might be used for these four exterior posts. My father, I remember, once cut elm trunks and allowed them to cure to be made into the four big central posts of a new lodge.

The central posts, if rotten at the foot, we pushed until they broke off. If they were not rotten, we cut them off with an axe. All this was done by women. Men helped only at lifting.

Exterior posts were pushed and broken off or felled with an axe, as were the main posts.

Posts that were rotten and not to be used in the new lodge were left to the owner for firewood.

The earth that I have said was scraped off the roof could be used again to cover the *atuti* puncheons of the new lodge. But many families preferred to use fresh earth again for the whole roof.

Clearing out stumps from postholes

As a new lodge was commonly built on the site of the wrecked lodge, it was necessary to raise the stumps of the old posts so that the postholes might be used again. The stump of a post was often pulled out by digging the ground away for six inches around the post. A stout ash pole (a) was rigged with a thong (b) bound in a notch at the end. The thong was made into a noose and was passed around the stump as low in the ground as possible.

Two women now took the ash pole and, rolling a log underneath to rest it on, they pried the pole as a lever.

At the same time, another woman with a stout ash stick pried at the stump on the opposite side.

To loosen it, in preparation for pulling it out, the woman had

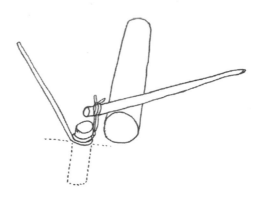

64. Post-pulling device. Drawing by Goodbird. (Courtesy of the American Museum of Natural History)

thumped the stump with an axe and had driven the axe blade into the center of the stump, splitting it slightly. The axe blade was then withdrawn. Thus loosened, the stump was soon pulled out.

An old lodge was wrecked in the spring often before the ground had thawed. When the new post was raised, coals were often put in the old posthole to thaw the ground so that the old post-hole could be cleaned and dug out for the new post. When a post had been broken off, and the rotten end was in the ground, fire was often used to thaw the ground so that the rotten end could be cleaned out.

The exterior posts of the new lodge were raised by the women. Two could raise a post with their bare hands.

Like-a-Fishhook Village and environs

Having lost approximately half their estimated population of four thousand after the catastrophic smallpox epidemic of 1781, the Hidatsas were thrown into chaos. Another smallpox epidemic in 1837 further reduced the population to approximately seven hundred (Stewart 2001: 329–331). The entire social structure was shattered—again. The Lakotas were established as the scourge of the Northern Plains and had already destroyed two of the three Hidatsa villages at the mouth of the Knife River in 1834 (Meyer 1977: 85). Many of the Hidatsa survivors were forced to move in with the Mandans.

In 1844 it was decided that the only way to survive was to move. The village of Like-a-Fishhook was established in 1845 on what, at that point, was the north bank of the Missouri River. Choosing a defensible terrace above the floodplain, the Hidatsas and Mandans also looked for abundant timber and land for their gardens. As recalled by Wolf Chief and Butterfly, there were seventy lodges in Like-a-Fishhook (Wilson 1934: 352), so the timber requirements were very high. The village was heavily fortified, and the trading posts at each end of the village further strengthened its position (Smith 1972). It was here that Buffalobird-woman and Wolf Chief grew up and Goodbird spent his first sixteen years.

Wolf Chief (vol. 10, 1911: 158–64; map from Wilson, *The Hidatsa Earthlodge* [1934]: 350)

Palisades of Like-a-Fishhook Village

Our village at Old Fort Berthold, as I recollect it, was surrounded by a ditch about 4 feet deep. The earth was thrown up in an embankment, and this was surmounted by palisades of logs,

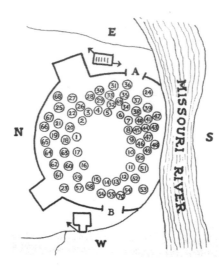

65. Like-a-Fishhook Village, on what is the north bank of the Missouri River at this point. The open area in the center was an accommodation for the Mandans, whose villages always had an open space for ceremonies and for the shrine for First Man. The "forts" on the east and west sides were trading posts. Adapted from a drawing by Goodbird. (Courtesy of the American Museum of Natural History)

some split, some round, about 8 inches or more in width, with 1.5 inches or so between so that we could shoot out. It had bastions, and at A and B there were two gates swinging on hinges [see figure 65]. The palisades were about 8 feet high. The fort was built a short time before I entered the Stone Hammer Society [the first of a series of age-graded societies for men]. Whitemen helped us build it. There were two forts on the east and west sides, and where you see the arrows, there was always a sentry. [These were trading "forts" or posts.]

Wolf Chief (vol. 13, 1913: 1183–84 or 229–30)

Old Fort Berthold Village, or Like-a-Fishhook Village as we called it, was palisaded in old times. The palisades were set in a ditch as close together as their natural shape permitted, and the ditch was then filled.

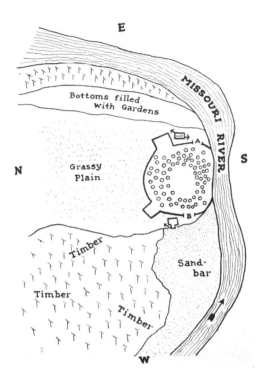

66. The site for Like-a-Fishhook was chosen for its defensible position and resources nearby. Note the abundance of timber. Map by Goodbird with instructions from Wolf Chief. (Courtesy of the American Museum of Natural History)

Dirt was then thrown up against the wall made by these palisades, or perhaps I should say against the fence made by these palisades, about 3 feet with a hollow ditch about 3 feet deep behind it. The ditch was on the inside of the fence. Loop holes were cut for muskets and arrows. The palisades were too high for enemies to climb.

At places along the wall or fence were placed bastions. These bastions projected outward beyond the line of the wall but were a part of the wall or fence itself.

A bastion was provided with a platform that was mounted by a small ladder. This platform was not very high and was made

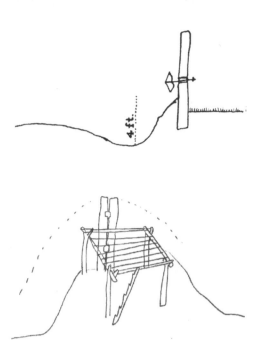

67. (*top*) The palisade, with a loophole and ditch inside, which offered further cover. Drawing by Goodbird. (Courtesy of the American Museum of Natural History)

68. (*bottom*) A bastion, a feature of many fortified villages in the area for centuries. The dotted lines, representing the palisade, are Goodbird's device for not cluttering up the features he wanted to show, including the loopholes through which defenders could shoot, the platform on which multiple individuals could stand, and the notched log ladder that provides access to the platform. (Courtesy of the American Museum of Natural History)

much like a garden watching stage. It rested upon four forked posts. Loop holes were then cut in the palisades, both above and below this platform, so that if the village was attacked, men might shoot out of these loop holes, some of them using the loop holes below this platform and some of them mounting the platform itself and shooting out of the loop holes above it.

SEVENTEEN

Miscellaneous Material

Basket making

Baskets were still being made and used at the time the Wilson brothers were at Fort Berthold, and Frederick did a detailed study of their manufacture, which included his trying his hand at making one. A historical account of the origins of Hidatsa (and Mandan and Arikara) basket making was published by Mary Jane Schneider in 1984.

Gilbert L. Wilson (vol. 18, 1915: 490)

Making of a bark basket

In the summer of 1912 I was on Fort Berthold reservation with my brother and hired Buffalobird-woman to teach us the weave of a bark basket. This was accomplished with some difficulty as she had been accustomed to expect special pay for instruction in old times. The year after, when I returned to the reservation, she wanted to know if I had been making any of the baskets for sale in Minneapolis!

Buffalobird-woman (vol. 25, 1912: frames 0084–0086)

Our Hidatsa name for basket was *midaxisi* or holder (frame 85).

Pay demanded for teaching basket making

Basket makers would not let others see how they worked, for if another wanted to learn how to make baskets she should pay a good price for being taught. All basket makers did thus. They would not teach another how to make baskets without being paid.

Another reason why I did not want people to watch me at my work (weaving baskets) was because they asked me questions. These questions bothered me so that I was apt to make a mistake with the splints. I did not like to have people around when I worked.

This custom (of not letting others watch us work) was also true of pottery makers.

When I was working at making a basket and someone hung around and watched me in order to learn how I did it, I would work too fast so that she could not see how it was done, because I did not want her to learn.

That is why there are so few basket makers on this reservation. They do not want to pay to learn, and we (basket makers) do not want to teach them when they do not want to pay for it.

More than one woman has thought she could learn to make a basket by just watching. This was just stealing, but it won't do any good because she cannot learn without a teacher.

We used these baskets to bear burdens on our backs, but their especial value was to carry corn and vegetables of our gardens to the village, especially the corn.

Three sizes of baskets made

We made three sizes of these woven bark baskets; you have purchased (examples of) all three kinds for the museum. This basket I am now making for you is the largest size that we used.

Use of baskets

We also carried our clay pots as well as our other cooking utensils in these baskets when we moved from our summer to our winter village. We often used baskets to bring snow into the house to melt for water.

In raising the basket full of corn, the thong or band was passed over the head and rested across both shoulders above the breasts. As the woman got to her feet (she had knelt), another helped her by taking the ends of the rear standards of the basket in either hand, and she raised the basket in place.

We did not fill the basket just loosely with corn. It was filled up to about three or four inches from the top when a row of ears was stood on end quite around the inside of the rim of the bas-

ket. Corn was then heaped within. These up-jutting ears prevented the corn from spilling out.

Both husked and unhusked corn was brought in baskets. The
green corn that we boiled to keep for winter was (brought home)
unhusked.

Buffalobird-woman interviewed by Frederick N. Wilson.
(vol. 12a, 1912: 62–68)

The wood used in the framework of a basket is of willow *maxoxica* [peachleaf willow–*Salix amygdaloides*]. This wood is light
and strong. The bark from this tree was also used for the black
splints or strips used in weaving the design, and it was colored
or dyed with clay.

The way that this was done was by taking the bark out to a
clay bed (slough or marshy places where alkali was formed) and
burying it. She took us to such a place and showed us one she
had in preparation. But mostly, as she says, in winter sage was
used to color bark black. She says that she takes some of the clay
that colors bark and puts it in a pot near the stove or fire, soft-
ens it with water, and buries the bark in this.

"It must be watched carefully not to spoil it—if left too long
the bark gets rough and rots. The length of time required to
color it depends pretty much on the weather—if good, it takes
less time—if not it takes longer, but the average time is about
seven days. The state of the bark is determined, however, more
by examination than by the length of time. I look at the bark,
and if it feels just right, take it out, as it spoils if left too long."

The white strands or strips are made from the bark of the boxelder [*Acer negundo*] (*mitetadiki*).

Sometimes baskets are made entirely of the willow bark. They
are brown and would last a long time.

Baskets should always be woven in the shade to prevent drying too fast. The best place was, like in old times, in the earthlodge, for there it is always cool and a little damp.

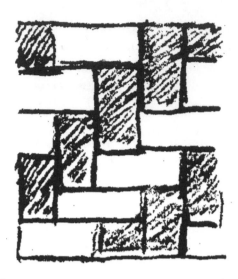

69. A basic basket pattern. Drawing by Goodbird. (Courtesy of the Minnesota Historical Society)

Baskets should never stand in the sun. If they get dry and warped, spray them all over with water and put them in a shady place.

The bark for baskets was taken from trees about the size of a man's wrist. If the tree was larger the bark was too heavy and if smaller the bark was too thin.

The frame parts were evidently selected "by the eye" for size and uniformity and were of size in relative proportion to the size of the basket wanted.

Buffalobird-woman interviewed by Frederick N. Wilson (vol. 12a, 1912: 65–66)

Observations on learning basketry

Among our people, when we wanted to learn to weave a basket, we had to pay to learn, for all those who were basket makers had the right to make them and to teach others and had to pay to learn in their turn. (This "right" constituted a monopoly and could be acquired only by purchase. FNW)

We always knew what to pay—it must be something valuable. In old times we paid a woman's robe and belt, a dress of blankets, or a dress of calico and something valuable as an ornament.

Sometimes when we wanted to learn we took a good tanned robe, went with it to the basket maker, and said, "You show me how to make a basket."

Then, if the basket maker did not have the materials on hand, she would take the pupil into the woods, show her the kind and size of trees, just what parts to use, and how to cut and prepare them. She would then show her how to make a basket. After this the pupil would go home and try to make one, and if she could not do it, she would come back and pay again, this time with dried meats and foods, a whole arm-full.

I do not know why it is that we do not teach everyone. Just that if you don't want to you don't. And anyhow, we wanted to keep the privilege to ourselves, as it gave us the right to get gifts, and if everyone learned and did not pay then we would not get anything.

(The inference is that either one had to buy baskets already made to order from the basket makers or else, if you wanted to make one, you had to pay another who had the "right" to teach you. FNW)

When I learned basket making I did not pay anything to learn as my teacher was a "relation" and stayed in our lodge. This was forty years ago, and I am now seventy-two years old. My teacher was Lone Woman. (My notes have the relationship thus: Lone Woman, Calf Woman's mother, is a sister by Indian custom. FNW)

I did not ask to learn, but Lone Woman thought she was getting old. So one day she asked me to go to the woods with her and help peel bark for baskets. Still she did not say anything about my learning, but when we came to the timber she showed me the right kind and size of tree for bark and how to peel off the bark for baskets.

When we came back we dried these barks or splints. Then she

took me out and showed me not only how to bury the willow splints in clay to color them black but also how to watch them to see that this is done right.

She told me about wetting and cutting the white barks or splints.

I learned in the earthlodge. This was the best place as the lodge was always cool and damp, and the barks would not dry out as quickly. Then too, we could be more out of the way and secluded and not so many people would see what we were doing.

If people did come in we just went off by ourselves in some other part of the lodge.

At the present time (1912) there are on the reservation about ten basket makers as nearly as I can recall. I will give their names and the kind of basket each makes:

On this side of the river at Independence:

Round Baskets

  1) Mikikash (Hides and Eats)

  2) Calf Woman

  3) Leader

Square Baskets

  1) Maxidiwiac (Buffalobird-woman)

  2) First Sprout (Her mother by Hidatsa convention)

  3) Pretty Woman

  4) Young Beaver

  5) Mikikash (Hides and Eats)

Across the river at Elbowoods:

Square Baskets

  1) Eagle Woman

  2) Paints Up

These are all old women, except one, and I do not know who their teachers were.

I asked her to name the parts of the basket, and they were given to me in this order, probably in the order of their importance.

1: *Aduidada*—Basket mouth

2: *Awaiadat*—The feet bows—the bow frames, especially the nether frames that rested on the ground.

3: *Itsesti*—The whole bottom part

4: *Aduikiikuha*—Strap side basket—the back

5: *Aduashakedetai*—Like horns—the parts of standards projecting above rim.

6: *Midukiake*—The basket itself was called *midukisha*, "pottery holder." Why she did not know.

Buffalobird-woman (vol. 25 n.d. [probably 1912]: frame 0083)

Willow bark baskets and splint designs

The white bark (splints) on this basket are of boxelder or *mitetadiki*, and the black splints are of *maxoxica* willow [*Salix amygdaloides*] bark. *Maxoxica* bark makes the stronger splints.

I once made a basket wholly of *maxoxica* willow bark and it lasted a long time. I did not color (dye) it, but it turned rather brown after a time. I made it for myself—it was not a pretty enough basket for a gift.

Other women sometimes made baskets like this. Such a basket was valueless to the owner except as a working basket for it was not handsome. Other women would say, "That basket is not pretty. I do not want to buy it."

But we knew that a *maxoxica* bark basket would last longer. The basket which I made for you was woven all over in diagonal patterns such as I use on the bottoms of all the baskets that I weave. I commonly wove the back of a basket in one design while the front and the two sides I wove in another or very similar design. Nowadays I sometimes weave each side of the basket differently to show the different designs that I can use.

The reason of our old time custom was that the back of the basket could not be seen because it rested against the shoulders of the owner. I therefore used an easily woven design on that part of the basket since it did not show anyhow.

Buffalobird-woman (vol. 10, 1911: 297)

Dice baskets, iron pots

We made small platters or flat baskets with a shallow edge for playing dice. A similar basket, but with a hole in the bottom, was made to use in the ceremony of the adoption of a sacred child.

Dice baskets were made from small willow shoots or, better yet, from the roots of the same plant. The baskets made of the roots are softer and more pliable. In the spring when the ice has left we can get these long roots easily along the riverbank.

In my time we had tin platters and iron pots, but we still used old-fashioned wooden bowls. Cooking done in a native clay pot tasted better. After iron came in we thought that food cooked in an iron pot did not taste so good. It would be passed along and one would smell of it and say, "This is cooked in iron," and not eat much of it. We could always tell when food was cooked in iron.

(Buffalo Bird Woman's complaint about a new iron pot is well founded. Wolf Chief states that copper and brass pots preceded iron, the latter coming in rather late. Goodbird adds of the willows that the long root of the kind of willow wanted for baskets was detected by its sweetish taste and that children chewed the roots sometimes for that reason. GLW)

Native drinks of the Hidatsas

Buffalobird-woman (vol. 11, 1912: 246–50
and vol. 27, 1912: 32–38)

We Hidatsas used a number of drinks in old times; at least we thought of them as drinks, although you may think of some of them as broths or soups. But we used them at meal times very much as you use your drinks.

Broth from marrow bones

We pounded and broke buffalo bones and boiled them for maybe half a day. We skimmed off the yellow bone-grease that rose to the top of the water and put it away. The water or broth that was left behind we thought very good to drink. We never wasted this liquid or threw it away. If we had too much for one meal, we carried the unused liquid over to the next meal or shared it with our neighbors. The name of this broth is *hidunahtihupa*, bone pounded broth.

This and other meat broths made by boiling both dried meat and fresh meat we drank at our regular meals. We had small wooden dishes in old times and spoons of buffalo horn, Rocky Mountain sheep horn, and mussel shells. When we passed the food around to anyone, he added broth enough in his dish, and when he ate the meat, he drank the broth. All these broths we drank out of a spoon.

Our old time cup was a Rocky Mountain sheep horn spoon. The clay pots which we made had rims too thick to be used as cups. For myself I did not care about drinking at meal time. Sometimes I took a little broth with my meat, but as a rule I ate my food without drinking. Many other people did likewise.

70. Wooden bowl with horn spoon. Drawing by Frederick N. Wilson. (Courtesy of the Minnesota Historical Society)

Teas

We did not use tea in old times, for tea and coffee were brought to us by whitemen, but as we got to know their use and learned also to use sugar, a number of native teas have come into use that we never made until we got sugar from the whites. There are a number of these teas we still use when we lack whiteman's tea or coffee.

Elm bark tea

The inner bark of the elm tree was removed and bound loosely in a ball and boiled while still green. We never did dry the bark before boiling. This boiling the bark made red tea. Although elm tea has come into use only since we obtained sugar, we used to chew the bark without boiling it for its sweetish juice.

Rose bark tea

We boiled the inner bark of rose bushes, first peeling off the red outer bark. The tea thus made tastes much like whiteman's tea, and this again we drank only after we obtained whiteman's sugar. We never chewed this bark as we did elm bark.

Mint tea

We also made a tea from mint leaves, and this tea we have used as a medicine from very old times. If a woman gave birth to a child and there was bleeding and the blood became thick, the mother was given mint tea to drink. This kept her body cool and made the blood run more freely so that the woman would not die. [It could be either *Mentha arvensis* L.—wild mint or *Monarda fistulosa* L.—wild bergamot. Hart (1992) lists *M. fistulosa* as being used for complications of childbirth.]

Boxelder sap

Another drink we had was boxelder sap, and we could get only a little of this in the spring. We broke a small branch off the tree four or five feet from the ground, bent the end of the stump downward, scarring the upper surface a little, and on the ground beneath placed a vessel to catch the sap drops. In a single night there would be collected perhaps two inches of sap in a small vessel. We were never able to collect much at a time.

The sap was boiled slightly, taken off the fire, and drunk at once. It was sweet like sugar. I have heard that the Chippewas make sugar of this sap, but I have never heard that my tribe ever did. We thought this sap drink very good, and we called it *mida-adahi* from *mida* wood and *adahi*, an old time word whose meaning I do not know.

Missouri water

But first of all drinks is water, and our best water is from the Missouri River. There are few springs in our country, and many of these are not good. The water often is full of alkali, and men who drink it get sick. Water from creeks is too warm in summer and is not good.

How our meals were served

Goodbird (vol. 18, 1915: 588–91)

When I was a small boy, and we all lived in Small Ankle's lodge, I remember we always ate our meals sitting around the fire. Between the two main posts nearest the door was my grandfather's couch [a wood platform], and between the two rear main posts was a long, low bench or stool. The members of the household sat on these two—the couch and bench—or on the floor.

Small Ankle sat with his wives, one of whom sat on the bench with him and the other on the floor. Wolf Chief and his wife, Coyote Woman, sat together next to them and then Son Of A Star and his wife, Buffalobird-woman (my father and mother), and myself. Next were two of my unmarried uncles, Full House and Flies Low, and finally Red Kettle and his wife, Young Beaver.

Our foods were on the floor before us in iron dishes about the size of a common washbasin.

My father, mother, and I used a single dish, which sat on the floor without a skin or mat beneath it. My grandfather, Small Ankle, and one wife used one dish, and his other wife used one dish. Wolf Chief and his wife ate from one dish. Flies Low, I remember, sometimes ate with his mother, Strikes Many Women.

I usually sat at one side of my father or mother but not between them.

Strikes Many Women, I remember, did most of the cooking, which was done in a three-legged kettle that we owned. When she was finished cooking and ready to serve, she would call, "*Mawu tats!*"(We eat) or "*Madu tada*" (You people eat). Then we all arose and took our places.

Strikes Many Women had the dishes all ready and piled before her. She would take the kettle off the fire and set it in front of her with the meat steaming and hot.

The hot pieces of meat she lifted from the kettle with a sharp stick. This stick she also used to stir up the cooking meat and get the top pieces to the bottom where they would cook quicker.

She had put a big wooden bowl beside her, and as she sat, she lifted all the pieces of meat out of the kettle into the wooden bowl. With a knife she then cut the pieces up into smaller sizes.

The steam from the meat pile would come rushing up into her face. I remember she would blow, "Whew," blowing the steam away as she cut off a piece of the meat with her knife and, lifting it by one corner with fingers and thumb, she laid the piece on a dish. She lifted it by the corner because it was hot. She probably also blew to cool the piece a little at the place where she was going to take hold of it.

We never fried buffalo steak. We boiled or roasted it. We ate meat every day. If we ran out of meat, the hunters usually went out and brought in more.

As Strikes Many Women filled each dish, she handed it to one of us boys—to Full House or me—saying, "Give this to so and so."

My grandmother, Strikes Many Women, served my father first, because he was her son-in-law, and she thought a great deal of him. Next came Wolf Chief, then Red Kettle, Full House, and Flies Low, and last, Small Ankle and his wife.

My grandmother did not speak my father's name but said to me, "Give this to your mother." Because he was her son-in-law she could not speak Son Of A Star's name, but she thought much of him nevertheless.

As the dish was brought up to him, my father took it and cut the meat into pieces for my mother and me as well as for himself. With his knife he would cut off a piece convenient to hold in one's hand and give it to me. I would bite mouthfuls from it, and when it was all eaten, I would ask for more.

When I was very small, my father or mother chewed up my meat for me.

71. Goodbird (*center*) eating with his father, Son Of A Star, and mother, Buffalobird-woman, in a hunting camp. Meat processed by Buffalobird-woman is being dried on the rack in the background. Drawing by Goodbird. (Courtesy of the Minnesota Historical Society)

Strikes Many Women, after taking out the meat, dipped up the broth left in the kettle into cups which she handed around. She dipped the cups directly into the kettle to get the broth. This was to drink. My father and mother each had a cup, I think, but I had none and drank from my father's and mother's cups.

If we were eating any boiled preparation of pounded corn such as succotash or mush, we ate with buffalo horn spoons. My father and mother each had a spoon, but I was fed by one of my parents.

Strikes Many Women, being the one who was serving, went without if the meal proved to be short for all who sat down. If there was an abundance, she helped herself to her share.

This was Indian custom. One who served always did this whether man or woman. Perhaps afterwards she would cook a portion for herself, but my grandmother always went without if the food was short.

After all had eaten, my mother collected the dishes. I do not remember what was done with them then.

Nettles (no use whatsoever except for practical jokes)

HIDATSA NAME: *maixake* (pimple-maker)
LOCAL ENGLISH NAME: (none given) (stinging nettle)
BOTANICAL NAME: *Urtica dioica* L. ssp. *gracilis* (Aiton) Seland or
Urtica dioica L.

Buffalobird-woman notes that these plants have red or purple blossoms, which probably means that they are *Urtica dioica* L. ssp. *gracilis* (Aiton) Seland. Flowers of *U. dioaca* are whitish or greenish. The two varieties completely overlap in range. The plants are obviously nettles given the description of the reaction they cause to the person coming into contact with them.

Buffalobird-woman (vol. 20, 1916: 339–41)

There is another plant, besides poison ivy, that we know is poisonous. We call it *maixake*. The plants grow about four or five feet tall. The have red or purple blossoms. They grow in the hills. If one took the leaf of this plant and whipped one's hand, it made him want to scratch the hit place, and the skin turned red and swelled up. But in a couple of hours the itching sensation and the redness were all gone.

Young men often would run against the plant just for fun and to make others laugh.

Now we Indians like to use the sweatbath, and when in the bath the bathers beat their bodies with little bunches of sage or other things. Once a man named Heart's Enemy was in a bath with some men friends. He had a bunch of chokecherry branchlets, and on the outside of the bunch he added a few of these *maixake* plants. With this bunch of branchlets he brushed himself very carefully and in such a way that the *maixake* plants did not touch him. Then he said to his friends, "I will fan you!" "That is good!" answered his friends. Heart's Enemy then struck his friends lightly on the back at a place where they could not reach to scratch. Soon he went out of the bath. Shortly after, all the others came hurrying out also, stinging and smarting where he had struck them, but they could not reach to scratch the place! Heart's Enemy did this just for a joke on his friends.

A man named Naxaxi Makue, or High Backbone, was one day sleeping in the camp, and the place was near a lot of *maixake* plants. Now these plants have leaves looking very much like the plant we call sweetleaf [hyssop].

High Backbone was naked except for his clout. Some of his friends for a joke went out and gathered some of the *maixake* leaves and lightly brushed his arms and legs as he slept. He awoke and began to scratch his skin, and he did so until he became ashamed and ran into the tent. When he came out again he was red all over.

He had not seen the young men when he awoke; he soon, however, knew what had happened!

The word *maixake* means "pimple-maker." The plant was put to no use whatsoever, and we avoided it with some fear.

(No specimen of this plant was sent in to the university for the reason that none grew near Independence. After Mr. Haupt left, on the way to Elbowoods, when driving thither with Goodbird, we found a patch of this plant. It was evidently a species of nettle. GLW)

[Haupt was the assistant sent with Wilson by University of Minnesota botany professor Josephine Tilden with whom the 1916 survey was to be a joint project. Apparently nothing ever came of it.]

FAMILY *Urticaceae*–nettle family
GENUS *Urtica* L.–nettle
SPECIES *Urtica dioica* L.–stinging nettle or
SUBSPECIES *Urtica dioica* L. *ssp. gracilis* (Aiton) Seland.–California nettle

Forest fire

Gilbert L. Wilson (vol. 9, 1910: 208)

The Indians have been much alarmed for the last two or three days. And today, Aug. 23, there is some suppressed excitement. There is evidently a forest fire somewhere west, probably in Montana, and the atmosphere is very murky and the sun obscured. Goodbird says that many of the women are much frightened and wonder whether the world is coming to an end. His wife, he adds, when she built a fire this morning, was made very uneasy because the flame, which should burn yellow, burned white. To show what this looked like, Goodbird made a small fire of a few chips which, at a distance, did burn with a white flame owing to the smoke with which the air seems filled. Goodbird says that he is not at all alarmed because he doesn't think the world is coming to an end yet. I must confess that the atmosphere does make things look a little ghostly. If I recollect correctly, the Legislature of Pennsylvania was quite as much alarmed during one of these dark days as are the women of the reservation. This phenomenon from forest fires and prairie fires is not uncommon in this latitude, but I have never seen the atmosphere so pronouncedly ghostly as today.

CONCLUSION

Hidatsa uses of plants were to a very large extent defined by the period during which they were being collected—in this case, from 1845, when they settled at Like-a-Fishhook, to Wilson's fieldwork between 1906 and 1918.

This was a period of calamitous change in the form of disease and intensive warfare with other groups but especially with the Lakotas and Assiniboines. Iron and brass were well established as necessities in their material culture, and both guns and horses were in widespread use at the time. Non-Indians became increasingly influential from the time of Lewis and Clark (1804–5) as more and more outsiders made their way into the formerly remote (to them) Upper Missouri region (the Middle Missouri of anthropologists and archaeologists).

Iron tools quickly replaced all stone tools with the exception of the stone hammers used to crush bones, pulverize dried meat, and mash berries. Copper, brass, and iron pots replaced pottery, although wooden trenchers continued to be used, and the wooden mortars and pestles were in use into the early twentieth century.

Gathering of plants lost most of its importance as substitutes became increasingly available. There seems to me to be a bit of nostalgia involved in the gathering of some plants—plants that were gathered because of the social aspect of gathering and, of course, the good taste. Picking juneberries would be a good example of this.

Other plants were gathered because they filled specific needs, such as plants used for medicine, twine, or basket splints. Some plants are still imbued with spirituality.

With the establishment of Like-a-Fishhook Village, traders became part of the community. Wolf Chief ran a store in Independence for years, and through these mechanisms a multitude of alien foods, spices (sugar being very important), tools, cooking equipment, guns, and what at the time were called "notions" made their way into the community. The trading posts, the military, missionaries, tourists (like Prince Maximilian and Audubon), and the continual contact with other people living on the plains or adjacent to them all served as vehicles for new ways of doing things and new things to contemplate.

The speed of culture change was immense. How much "traditional" Hidatsa was lost when perhaps as much as 80 to 90 percent of the population died as a result of disease over a period of less than sixty years? The remaining population had to do whatever it took in order to survive, even if it meant, for example, bypassing marriage conventions, which at one time were regulated by clan membership and were virtually inviolable.

Despite the fact that much of Hidatsa culture was shared by all, a lot was specific to individuals who had rights to that information or societies who held the information. When these individuals died or entire groups died, so did the information.

And keep in mind that Wilson had never intended to do a thorough job of gathering information about the Hidatsa uses of plants but took on that task in addition to the work he had planned to do.

His own philosophy of doing ethnography was not to structure his interviews other than to keep the topic being discussed more or less on track. Thus the information collected is very uneven, and the "right questions" were sometimes not constructed, let alone asked, because he wanted his sources to talk about what they wanted to talk about.

The series of epidemics that swept through the Western Hemisphere killing what were probably tens of millions of people (see Denevan 1992 for a variety of views on pre-European contact) greatly impacted the Hidatsas and their neighbors (and originally cautious allies), the Mandans. The epidemics of 1780 and 1837 reduced the Mandans to fewer than 150 people and the Hidatsa by at least a third by the time of Buffalobird-woman's birth in 1839 (see Meyer 1977: 90–107) for the conflicting historical accounts of populations and losses from the smallpox and other epidemics. Despite the chaos caused by losing much of their populations, the survivors managed to develop a new society in which many of the old rules no longer made sense and much traditional information was lost as the individuals who knew important parts of the culture died suddenly before that information could be passed on.

Thus Gilbert Wilson's sources, including Buffalobird-woman, were survivors who grew up in a time of incredible cultural chaos and yet established new norms to fit the existing and constantly changing realities. Citing the census of 1910 Lowie noted that there were 547 Hidatsa remaining (1917: 17). They switched from stone tools to iron tools, from pottery to brass and iron cooking wares, from earthlodges to log cabins, from compact villages to individual homesteads, from leather clothing to cloth, and from the Hidatsa religion to Christianity. Buffalobird-woman retained her Hidatsa beliefs, her brother Wolf Chief made a midlife conversion to Christianity, and her son Goodbird became a clergyman.

Wilson gathered what he could during the twelve summers that he visited Fort Berthold Reservation from 1906 to 1918. Reservation life itself required an entirely new way of living and accommodating to the often arbitrary impositions of both the government and the newly arriving hordes of settlers. This changed again in 1888 when the federal government decided that the last earthlodge village of Like-a-Fishhook, into which both the Mandans and Hidatsas had moved in 1845 (and to which yet another group of survivors, the Arikaras, had moved twenty years later), would be abandoned and its inhabitants relocated on individual allotments of land. This plan was supposed by the government to be the best way to integrate Indians into American society.

The results have been decidedly mixed. The fact remains that the reservations still exist and are each cultural centers where ethnicities persist and evolve. Transitions from one generation to the next have not been easy, as economic forces completely out of their control have compelled the Hidatsas and all Indians to realign every aspect of their cultures. Political decisions made with little or no consultation with the people whose lives were to be affected created even greater confusion.

Plants introduced both accidentally and purposely by Europeans and Americans are now included in modern ethnobotanies as well as some of the plants used for as long as there are oral traditions and archaeological records. This book records some of the knowledge of plants known to Hidatsas who lived in the last earthlodge villages. Wilson fortuitously arrived just in time to record a great deal of middle to late nineteenth- and early twentieth-century Hidatsa life as recalled by those who had lived it.

For greater insight see the chapter on "Family Ties" in *The Way to Independence* by Carolyn Gilman and Mary Jane Schneider (1987: 20–21) for a good, brief explanation of Hidatsa social organization. This book is also the most complete compendium of Hidatsa material culture and contains details of the lives of those who worked for so many years with Gilbert Wilson. Detailed accounts of social organization can be found in Robert Lowie's *Notes on the Social Organization and Customs of the Mandan, Hidatsa, and Crow Indians* (1917) and Alfred Bowers's *Hidatsa Social and Ceremonial Organization* (1965). Roy Meyer's *The Village Indians of the Upper Missouri* (1977) is a thorough history of the period.

This collection of memories of plants and their uses is the best there is, and one wishes that there were more. Some missing information can be guessed at, but anyone who has read ethnobotanical material for a region knows that a plant may have been used in different ways by different people or peoples. Any information collected at present is just that—contemporary ethnobotany— interesting in its own right but not to be projected backward without considerable trepidation.

By using Buffalobird-woman and Wolf Chief's information and anecdotes, Goodbird's translations, and Wilson's transcriptions, we get as close as we can to a past now roughly five generations away at the least.

APPENDIX

Frederick N. Wilson's Comments on "The Hidatsa Earthlodge"

Gilbert Wilson died in 1930 with much of his work unpublished. Over the years at Fort Berthold he and his brother Frederick had collected a great deal of information on earthlodges, particularly from Buffalobird-woman. Bella Weitzner of the American Museum of Natural History compiled and edited the Wilsons' material for the museum, which published it in 1934. Prior to publication the AMNH sent the proofs of the finished manuscript to Frederick Wilson for comment. He had worked closely with Gilbert and had drawn quite precise "blueprints" of several lodges as well as details of their construction. He thus knew more than anyone the details of Hidatsa earthlodge structure and construction. The AMNH seems to have paid no attention to his comments and corrections at all. Because his observations and criticisms are, I think, significant, I have included them here.

Notes and Comments on the Galley-proofs of the Manuscript of *The Hidatsa Earthlodge* by Gilbert L. Wilson
Arranged and Edited by Bella Weitzner

After reading this paper, there seemed to be so many conflicting and contradictory statements to account for or to reconcile that I hardly know where to begin.

I shall make no attempt at a critical analysis, as that is beyond me; and as far as arranging and coordinating the material is concerned, Miss Weitzner deserves the highest praise, as I can understand the terrific labor involved. There is however a certain feeling of uncertainty or confusion arising from all these apparently irreconcilable statements, and I think a little additional matter might be put in at the proper place to obviate this, as it seems to me that these discrepancies are more seeming than actual.

Before going into any detailed suggestions, I should like to make a few general remarks which may be taken for what they are worth.

The older Indians had no great familiarity with precision instruments, and so could not be expected to be very accurate in giving measures from memory, and these measures and numbers should be taken as approximate, rather than as precise, in many instances. Familiar as we are with measures, most of us are hopelessly inaccurate in estimating measures and distances.

As to accounts of the early travelers, from my own experience I can say that appearances are often deceitful, and that many of the dimensions and details of lodge construction were accepted only after making actual measurements.

After looking over all the source material at hand, and many early accounts, we must, I think, accept the twelve post conical lodge, of about 40 ft. diameter as the norm, and all others however common as special types or variations.

Certainly the whole subject becomes clearer, if this is done, and any departure from this becomes merely a matter of individual preference, or possibly in the case of the flat-roof lodge, a matter of custom, for keepers of Ceremonies and for dances.

Both Buffalo-bird-woman and Hairy Coat gave me the distinct impression that, while there were many flat-roof lodges in the villages, they were not so numerous as to be common, and that, while there was apparently no restrictions as to their use, and any one might build them still those that had them were more apt to be among the most influential members of the community. These were more likely to be the Chiefs, the Keepers of Ceremonies, or any important and influential member of the tribe.

Just as with his more advanced brother the whiteman, the location of the lodge and its size and finish no doubt did much to proclaim his position and importance.

In comparing several early accounts (the few available to me), as well as the drawings and prints, the conical lodge is everywhere in evidence, and is, even in the often inaccurate descriptions and drawings, quite unmistakable. This convinces me that it is the normal type, and that the flat-roof lodge is a further improvement or development of this, different only in the smoke-hole construction and arrangement; in all other ways conforming to the same design.

To take an example from our own region here, we might liken the winter lodge and its associate "twin" lodge to our summer cottage at the lake; the conical lodge to our permanent home, and flat-roof lodge to the mansion of our society folk, or to a club house. This is not to be taken as a definite distinction, but as a suggestion.

To proceed to the galleys: In galley 3, there is a section which I think would be better if set over into the (can't make it out) section, as it is confusing here.

In the footnotes is again the discussion of excavating the floor of the lodge.

I asked particularly about this, but none of our informants would agree that it was done, although Hairy Coat said some people might have done so, but that he never knew it to be done. There was no evidence of it in any lodge examined by us, although going into a lodge does give one something of that impression. The earth in the entranceway is quite worn down; then one steps over a sill and again the ground is worn away so that one could easily get that illusion. As they preferred the ground to be of a certain kind, they may have dug down to remove any sandy or other unsuitable soil. This is suggested in the statement by Mathews.

I have never seen any statement regarding one possibility, but it seems to me to be entirely possible, that where a new village was located upon suitable ground, they might very likely have excavated, to save the labor of carrying the roofing earth for considerable distances, as it is evident that the ground area of the lodge is almost exactly equal to that of the roof.

Most of these statements about this matter are, I think, little more than assumptions, as most of the travelers expected to see a "dug-out" and therefore did.

Most of the families retained their assigned places in a village, and if the village remained any length of time in one place, each lodge would no doubt be rebuilt several times, this would in due course result in a condition similar to excavation. Not-a-woman speaks of making the renewal posts longer if the ground is much worn away within the lodge; this, taken together with the fact that the earth on the lodge is pulled down all around it in rebuilding, would soon account for such a seeming condition. I can not believe that all of this old earth was either carried away or all used again. If we assume this to be true, and then consider that the lodges were quite close together, and that there was constant washing down of earth from each lodge, this so called excavation could easily be accounted for.

We inspected several sites, especially the one at Knife river, and while the lodge circles gave this effect, one must remember that the greatest amount of earth was on this portion (the outer) of the lodge and would give the impression of higher ground without and a depression within the circle, when the lodge fell in.

Continuing on galley 3, regarding the posts for the lodge, I think this should be made a more general statement, as the posts were pretty well proportioned to the size of the lodge, and all the 8 central timbers were close to the same diameter, and this diameter had a rough relation to its length. I should not say that it was either with knowledge or intention that this came about, but the fact remains that the diameter in inches closely approximated the length in feet, of each timber. In general, a supporting member with ten feet in the clear, would be about ten inches thick; one twelve feet, clear, twelve inches, etc. For larger lodges, this proportion might diminish gradually, as timbers beyond 16 or 18 in. diameter would be almost unmanageable, except for uprights, when it came to a lodge using 20 ft. timbers.

The 8 central posts and stringers (beams), formed an almost exact cube, and the outer or atuti posts were almost exactly half the height of the completed central structure; when the beams were laid on these, the roof poles were at such a pitch as to give good drainage, yet not so steep that the poles were apt to slide off.

In the flat-roof lodge all the supporting timbers were raised relatively higher; this made it less apt to be smoky, made a better appearance, and made a much roomier lodge: more space could be used. This would also partly account for its use as a Society lodge.

In galley 4, mention is made of the lower stringer in the central group being long and heavy. This was only done in the ordinary lodge to allow for squaring the ends (i.e. not to weaken it), but was probably true for the larger lodges, and more especially the flat-roof, as here there were more timbers and greater weight.

It sometimes happened that a family would want more room in a lodge, and then they would build the lodge with a Tail to it. This was done by using a heavier beam or stringer and extending it a foot or so beyond the rear post on which it rests.

The rest of the central framework was then completed just as always. If the Tail was not used, the rear central posts were set far enough back to come under the end of the beam. (See drawings) The atuti measurements were then accommodated to this change, adding mostly to the length at the sides of the lodge.

Galley 5, discussing the flat-roof smokehole, gives an erroneous impression, for both Buffalo-bird-woman and Hairy Coat mentioned the use of the slabs or planks and then of willows, grass and earth, all at the same time. My understanding was that this was customary and that any other way was an individual variation.

The smokehole planks were also carried around all four sides—all the models made for us showed this. (See drawings)

The grass used, according to Hairy Coat, was the long slough or a marsh grass, as that was said to be tough.

Galley 6—Relative to the depth of earth on the lodges, I do not think we measured any that had a greater depth than 5 inches. We did not measure all, and those were not in the best of repair, but there was not much difference apparent in the depth on any of the lodges, except those falling in. A depth of a foot or so as stated by some would be a terrific weight, even though the base was sod, which was the usual way. I know we thought the depth was greater until we measured, as the rafter, willows, grass, and earth gave a massive feeling to the roof.

Galley 8—In describing the bed and its canopy or case, I think there has been a mistake made here. The bed and covering are properly described, but there is nothing in my notes about its having been bound to the rafters. If the word rafter refers to the framework of the bed, it would be better to so indicate. There was an extra canopy or hanging, that was attached to the Atiduti-du pole above and forward of the bed; this was drawn under the atuti stringer and over the bed, and allowed to hang back of the bed or else was attached to the atuti slabs. This gave protection against leaks from the roof, and greater warmth.

Galley 13—Under the heading "The Central Posts and Beams" there is the statement that the four forked central posts were about a foot in diameter, and about six feet high. As the rest of the text goes on to describe the average sized lodge, this is obviously wrong. It should read ten or twelve feet. All the other dimensions are given as of about the size of Wolf-chief's lodge. This, I was told by both of my informants, was close to the usual size of the large winter lodges, that is, usually built for a large family. This is undoubtedly true; otherwise horses could hardly be kept in them, too.

As this text goes over into Galley 14, and goes on to a description of the Twin or like-a-twin lodge, I shall take up the two matters at once. There is no need to waste a lot of words about the large winter lodge and its twin, for I have made a series of drawings showing that structure and relations just as described to me by Hairy-coat and Buffalo-bird-woman, and those I am sending to you. The like-a-twin lodge was added to the regular winter lodge by families where there were old people or as an extra warming room for families that could afford them. By referring to the drawings one can see that they must have been quite cozy, the angles of the roof poles being such as to set much on the principle of portable reflecting ovens or "bakers" used by woodsmen and campers. (Can't think of the name!) Hairy-coat gave me to understand that this so-called "twin" lodge was also used alone by small or poorer families, and that then it had an entrance on the same as the regular lodge.

In the series of photos of Owl Woman's lodge, taken for my brother in 1916 and 1917 by Mr. Page, can be seen an authentic "twin" lodge as it would be used thus, and looking exactly as I have pictured it from descriptions, except that I have, of course, drawn it as attached to the lodge. All descriptions are on the drawings.

In the remaining Galleys there is nothing I can add to, except that in 17 there is the matter of the Hidatsa tent tie. I send along some drawings that were made not so long after I returned from the Reservation, and I think are correct, but as I did not take down any text on this, can only send them along as being correct as far as I have recollection. This tie was made for us and I diagrammed it on the spot. So-o-o.

There are two or three specific recommendations that I should like to make which can be done without any rewriting, or without much of any change of place in the text. These are as follows:

1st—I think the main discussion of the making or providing a payment to the lodge builder and a feast to the other workers, ought to be taken up at the very beginning of the paper, in the building of the summer lodge. This was such an important thing and so associated with the whole subject, that we should establish this at once, and then refer to again, where called for, with such changes made or differences noted as the circumstances require. This can be done by almost lifting this part from the winter lodge account.

2nd—The material dealing with removal of posts etc., on the site of an old lodge would be better placed in its original setting, where the wrecking and rebuilding are taken in order. If this part is removed from Galley 3, it will give much more connected feeling to narrative. It is hardly consistent to wreck the structure before it has been built, and this gave me quite a confused feeling in first reading the paper.

3rd—The drying stage while detached, was considered an integral part of the lodge and was the scene of quite an important part of the family life. It was also such a distinguishing part of the village picture that it can hardly be left out consistently.

A few paragraphs devoted to this could be revamped from my brother's paper on Hidatsa Agriculture. I do not have the drawings but the cuts could be taken from book illustrations in the book.
   These are all the suggestions!

Frederick N. Wilson [undated but probably 1933]
(Source: Minnesota Historical Society,
Gilbert L. Wilson Archive)

BIBLIOGRAPHY

Ahler, Stanley A., and Phil R. Geib. 2007. Investigations at Double Ditch Village, a Traditional Mandan Settlement. In *Seeking Our Past: An Introduction to North American Archaeology*, edited by Sarah W. Nesius and G. Timothy Gross, 442–51. New York: Oxford University Press.

Ahler, Stanley A., Thomas D. Thiessen, and Michael K. Trimble. 1991. *People of the Willows: The Prehistory and Early History of the Hidatsa Indians*. Grand Forks: University of North Dakota Press.

Ashworth, Allan. 1999. Climate Change in North Dakota since the Last Glaciation: Review of the Paleontological Record. *Proceedings of the North Dakota Academy of Science* 53: 171–76.

Barkley, T. M., ed. 1977. *Atlas of the Flora of the Great Plains*. Ames: Iowa State University Press.

Blake, Leonard W. 1981. Early Acceptance of Watermelon by Indians of the United States. *Journal of Ethnobiology* 1(2): 193–99.

Blake, Leonard W., and Hugh C. Cutler. 2001. *Plants from the Past*. Tuscaloosa: University of Alabama Press. Originally printed in 1973 and a slightly expanded version released in 1976.

Bowers, Alfred W. 1965. *Hidatsa Social and Ceremonial Organization*. Smithsonian Institution Bureau of American Ethnology, Bulletin 194. Washington: Government Printing Office.

Brown, William L., and Edgar Anderson. 1947. The Northern Flint Corns. *Annals of the Missouri Botanical Garden* 34(1): 1–29.

Buffalo Bird Woman and Gilbert L. Wilson. 1987. *Buffalo Bird Woman's Garden*. St. Paul: Minnesota Historical Society Press. Also at http://digital.library.upenn.edu/women/buffalo/garden/garden.html (accessed August 2012).

Catlin, George. 1973. *Letters and Notes on the Manners, Customs, and Conditions of the North American Indians*. Vol. 2. New York: Dover.

Chagnon, Napoleon. 1997. *Yanomamo*, 5th ed., edited by George and Louise Spindler. Case Studies in Cultural Anthropology. Fort Worth: Harcourt Brace College.

Clark, William. 1804. *The Journals of Meriwether Lewis and William Clark*. 2 August. http://www.infoplease.com/t/hist/lewis-clark-journal/day81.html#ixzz2GrmTtjOW (accessed 13 January 2013).

Cook, E. R., D. M. Meko, D. W. Stahle, and M. K. Cleaveland. 1999. Reconstruction of Past Drought across the Coterminous United States from a Network of Climatically Sensitive Tree-Ring Data. In *North American Drought in Paleo Perspective: The Last 500 Years*. www.ncdc.noaa.gov/paleo/pdsiyear.html (accessed December 2009).

Cook, E. R., D. M. Meko, D. W. Stahle, and M. K. Cleaveland, with E. P. Gille, J. T. Overpeck, T. R. Karl, and N. B. Guttman. 2010. North American Drought Variability. www.ncdc.noaa.gov/paleo/drought/animation/pdsi_animation.html. In NOAA *Paleoclimatology*. www.ncdc.noaa.gov/paleo/paleo.html (accessed July 2011).

De Candolle, Alphonse. 1959. *Origin of Cultivated Plants*. Reprint of 1886 2nd ed. New York: Hafner.

DeMallie, Raymond J., ed. 2001. *Handbook of North American Indians: Plains*. Vol. 13, pts. 1 and 2. Washington: Smithsonian Institution.

Denevan, William, ed. 1992. *The Native Population of the Americas in 1492*. Madison: University of Wisconsin Press.

Doebley, John F., Major M. Goodman, and Charles W. Stuber. 1986. Exceptional Genetic Divergence of Northern Flint Corn. *American Journal of Botany* 73: 64–69.

Fawcett. William B., Jr. 1988. Changing Prehistoric Settlement along the Middle Missouri River: Timber Depletion and Historical Context. *Plains Anthropologist* 33(119): 67–94.

Galinat, W. C., and J. H. Gunnerson. 1963. *Spread of Eight-Rowed Maize from the Prehistoric Southwest*. Botanical Museum Leaflets, Harvard University, 20: 117–60.

Gardner, Robert. 1964. *Dead Birds*. Watertown MA: Documentary Educational Resources.

Gilman, Carolyn, and Mary Jane Schneider. 1987. *The Way to Independence: Memories of a Hidatsa Indian Family, 1840–1920. Publications of the Museum Exhibit Series 3*. St. Paul: *Minnesota Historical Society*.

Gilmore, Melvin R. 1919. *Uses of Plants by the Indians of the Missouri River Region. 33rd Annual Report of the Bureau of American Ethnology*. Washington: Smithsonian Institution; reprint, Lincoln: Bison Books, University of Nebraska Press, 1977.

Griffin, David E. 1977. Timber Procurement and Village Location in the Middle Missouri Subarea. *Plains Anthropologist Memoir 13*: 177–85.

Grinnell, George Bird. 1905. Some Cheyenne Plant Medicines. *American Anthropologist* n.s., 7: 37–43.

Hart, Jeff. 1992. *Montana Native Plants and Early Peoples*. Helena: Montana Historical Society Press.

High Plains Regional Climate Center. 2012. New Town 4 W, North Dakota (32383). *Period of Monthly Climate Summary*. http://www.hprcc.unl.edu/cgi-bin/cli_perl_lib/cliMAIN.pl?nd6383 (accessed June 2012).

Howard, James H. 1960. Butterfly's Mandan Winter Count: 1833–1876. *Ethnohistory* 7(1): 28–43.

Jenks, Albert E. 1901. *The Wild Rice Gatherers of the Upper Lakes: A Study in American Primitive Economics*. Extract from Nineteenth Annual Report of the Bureau of American Ethnology. Washington: Government Printing Office.

Johnson, Alex. 1970. Blackfoot Utilization of the Flora of the Northwestern Great Plains. *Economic Botany* 24: 301–24.

KellerLynn, Katie. 2007. Theodore Roosevelt National Park Geologic Resource Evaluation Report. *Natural Resource Report* NPS/NRPC/GRD/NRR. Denver: National Park Service.

Kindscher, Kelly. 1987. *Edible Wild Plants of the Prairie*. Lawrence: University Press of Kansas.

———. 1992. *Medicinal Plants of the Prairie*. Lawrence: University Press of Kansas.

Knapp, Alan K., John M. Blair, John M. Briggs, Scott L. Collins, David C. Hartnett, Loretta C. Johnson, and E. Gene Towne. 1999. The Keystone Role of Bison in North American Tallgrass Prairie. *Bioscience* 49(1): 39–50.

Laird, K. R., S. C. Fritz, K. A. Maasch, and B. F. Cumming. 1996. Greater Drought Intensity and Frequency before AD 1200 in the Northern Great Plains, U.S.A. *Nature* 384: 552–54.

Lehmer, Donald. 1971a. Personal communication.

———. 1971b. Introduction to Middle Missouri Archaeology. *Anthropological Papers 1*. Washington: National Park Service.

Lowie, Robert H. 1917. *Notes on the Social Organization and Customs of the Mandan, Hidatsa, and Crow Indians*. Anthropological Papers of the American Museum of Natural History, vol. 21, pt. 1. New York: American Museum Press.

Lucero, M. E., E. L. Fredrickson, R. E. Estell, A. A. Morrison, and D. B. Richman. 2006. Volatile Composition of *Gutierrezia sarothrae* (Broom Snakeweed) as Determined by Steam Distillation and Solid Phase Microextraction. *Journal of Essential Oil Research* 18: 121–25.

Lutz, H. J. 1958. Observations on "Diamond Willow," with Particular Reference to Its Occurrence in Alaska. *American Midland Naturalist* 60: 176–85.

Mac, M. J., P. A. Opler, C. E. Puckett Haecker, and P. D. Doran. 1998. *Status and Trends of the Nation's Biological Resources.* Vol. 2. Reston VA: Department of the Interior, U.S. Geological Survey.

Maynard, David, and Donald N. Maynard. 2000. Cucumbers, Melons, and Watermelons. In *The Cambridge World History of Food*, edited by Kenneth F. Kiple and Kriemhild Conee Ornelas. Cambridge: Cambridge University Press.

Meyer, Roy W. 1977. *The Village Indians of the Upper Missouri.* Lincoln: University of Nebraska Press.

Moerman, D. E. 1998. *Native American Ethnobotany.* Portland OR: Timber Press.

————. 2009–11. *Native American Ethnobotany: A Database of Foods, Drugs, Dyes, and Fibers of Native American Peoples, Derived from Plants.* Dearborn: University of Michigan. http://herb.umd.umich.edu/ (accessed September 2011).

Munson, Wendy. 2004. Interactions and Dynamics of a Native American Horticultural System: An Ecosystematic Approach. Master's thesis, Minnesota State University, Mankato.

Munson Scullin, Wendy, and Michael Scullin. 2005. Potential Productivity of Midwestern Native American Gardens. *Plains Anthropologist* 50(193): 9–21.

National Museum of Natural History. 2012. *Swift Fox (Vulpes velox).* http://www.mnh.si.edumna/image_info.cfm?species_id-419. *Kit Fox (Vulpes macrotis).* http://www.mnh.si.edumna/image_info.cfm?species_id-428 (accessed October 2012).

National Weather Service. 2011. *Williston ND Accumulated Yearly Precipitation.* Bismarck ND Weather Forecast Office. http://www.crh.noaa .gov/bis/ (accessed July 2012).

Penn Veterinary Medicine. 2011. Genus: *Equisetum.* http://research.vet .upen.edu/PoisonousPlantsofPA/Equisetumarvense/tabid/5438 /Default.aspx (accessed October 2011). Pennsylvania State University, Philadelphia.

Pike, Kenneth. 1967. *Language in Relation to a Unified Theory of the Structure of Human Behavior.* 2nd ed. The Hague: Mouton.

Potter, L. D., and D. L. Green. 1964. Ecology of a Northeastern Stand of *Pinus flexilis*. *Ecology* 45: 866–68.

Schneider, May Jane. 1984. An Investigation into the Origin of Arikara, Hidatsa, and Mandan Twilled Basketry. *Plains Anthropologist* 29(106): 265–76.

Scullin, Michael. 1989. The Adaptive Features of Tillering in Northern Flint Corn. *Plains Anthropologist* 34(123): 35–42.

———. 2005. Confounding Stereotypes: Building an Earthlodge for Fun and Edification. In *Plains Earthlodges: Ethnographic and Archaeological Perspectives*, edited by Donna C. Roper and Elizabeth P. Pauls. Tuscaloosa: University of Alabama Press.

Slusher, John P. 1985. *Wood Fuel for Heating*. MU Guide, G 5450. University Extension, University of Missouri–Columbia.

Smith, Bruce. 2006. Eastern North America as an Independent Center of Plant Domestication. *Proceedings of the National Academy of Sciences* 103(33): 1223–28.

———. 2007. *Rivers of Change: Essays on Early Agriculture in Eastern North America*. Rev. ed. Tuscaloosa: University of Alabama Press.

Smith, G. Hubert. 1972. *Like-a-Fishhook Village and Fort Berthold, Garrison Reservoir, North Dakota*. Anthropological Papers 2. Washington: National Park Service.

Stevens, Ora Alva. 1963. *Handbook of North Dakota Plants*. Fargo: North Dakota Institute for Regional Studies.

Stewart, Frank Henderson. 2001. Hidatsa. In *Handbook of North American Indians: Plains*, edited by William C. Sturtevant, vol. 13, pt. 1 of 2: 329–48. Washington: Smithsonian Institution.

Swenson, Fern E. 2007. Settlement Plans of Traditional Mandan Villages at Heart River. In *Plains Village Archaeology: Bison Hunting Farmers in the Central and Northern Plains*, edited by Stanley A. Ahler and Marvin Kay, 239–58. Salt Lake City: University of Utah Press.

United States Department of Agriculture. 2013. The PLANTS Database (http://plants.usda.gov). Baton Rouge: National Plant Data Center. Louisiana (accessed November 2012).

Upham, Steadman, Richard S. MacNeish, Walton C. Galinat, and Christopher M. Stevenson. 1987. Evidence Concerning the Origin of Maiz de Ocho. *American Anthropologist* 89: 410–19.

Wilson, Gilbert L. 1913. Letter to Clark Wissler. 6 May. Anthropology Archives, American Museum of Natural History.

————. 1916. Letter to Clark Wissler. 8 July. *Anthropology Archives, American Museum of Natural History.*

————. 1917a. Agriculture of the Hidatsa Indians: An Indian Interpretation. Studies in Social Sciences 9. *Bulletin of the University of Minnesota.*

————. 1917b. Letter to Clark Wissler. 8 June. Anthropology Archives, American Museum of Natural History.

————. 1923. Letter to Clark Wissler. 12 April. Anthropology Archives, American Museum of Natural History.

————. 1924. The Horse and the Dog in Hidatsa Culture. *Anthropological Papers of the American Museum of Natural History* 15 (2): 125–311.

————. 1934. The Hidatsa Earthlodge. Arranged and edited by Bella Wietzner. *Anthropological Papers of the American Museum of Natural History* 33(5): 341–420.

Wissler, Clark 1916. Letter to Gilbert Wilson. 3 July. *Anthropology Archives, American Museum of Natural History.*

Wood, W. Raymond. 1980. Plains Trade in Prehistoric and Protohistoric Intertribal Relations. In *Anthropology on the Great Plains,* edited by W. Raymond Wood and Margot Liberty. Lincoln: University of Nebraska Press.

Woolworth, Alan R. 1987. "Contributions of the Wilsons to the Study of the Hidatsa." In *The Way to Independence: Memories of a Hidatsa Indian Family, 1840–1920,* edited by Carolyn Gilman and Mary Jane Schneider. Museum Exhibit Series 3. St. Paul: Minnesota Historical Society.